QUANTUM CASCADE LASERS

Edited by **Vasilios N. Stavrou**

Quantum Cascade Lasers

http://dx.doi.org/10.5772/62674
Edited by Vasilios N. Stavrou

Contributors

Tsungtse Lin, Arpan Deyasi, Ning Yang, Yan Xie, Suqing Duan, Weidong Chu, Jian Wang, Yanfang Li, Hercules Simos, Adonis Bogris, Dimitris Syvridis, Raúl Pecharromán-Gallego

Published by InTech

Janeza Trdine 9, 51000 Rijeka, Croatia

Publishing Process Manager Romina Skomersic
Technical Editor SPi Global
Cover InTech Design team

First published March, 2017
Printed in Croatia
Legal deposit, Croatia: National and University Library in Zagreb

Additional hard copies can be obtained from orders@intechopen.com

Quantum Cascade Lasers, Edited by Vasilios N. Stavrou
p. cm.
Print ISBN 978-953-51-3061-1
Online ISBN 978-953-51-3062-8

Contents

Preface

Quantum cascade lasers (QCLs) are semiconductor laser structures, the emission range of which varies from mid-infrared range to far-infrared range of the electromagnetic spectrum. These lasers are constructed by a repeated stack of multiple quantum wells and operate on interband transitions of the semiconductor structures. These types of lasers have been commonly used in spectroscopy, remote sensing, optoelectronics, and communications. Due to their potential use in research and industry, advanced growth techniques such as molecular beam epitaxy (MBE), metal-organic vapor phase epitaxy (MOVPE), and metal-organic chemical vapor deposition (MOCVD), among others, have been used to fabricate qualitative quantum-cascade laser structures.

Topics related to the essentials of QCL structures, to the electronic structure of QCLs, to computational methods, to fabrication techniques, and to techniques in order to achieve low-noise operation with quantum-cascade lasers and their applications are included in the current chapters' collection. This book is divided into two sections. More specifically, in Section 1, the calculations on the band structure of cascade lasers and on the electronic band structure of QCL (e.g., transfer matrix technique, finite element method, and variational method, among others) under the existence of an applied electric field are presented. Moreover, a detailed investigation of electrical and optoelectronic properties of multiple-quantum-well structure is carried out. Furthermore, a theoretical study on the intensity noise characteristics of quantum-cascade lasers under the external noncoherent optical injection is carried out. This section ends with an important description on growth methods for QCLs and the possibility of developing GaN-based QCLs among other materials. A few applications of cascade laser structures such as power amplification and THz applications are presented, among others, in Section 2. The THz QCL design and fabrication toward the high-temperature and large-average output power operations for the real applications are described. Furthermore, the importance of the power amplification and coherent combination techniques to improve the output power maintaining the single-mode operation is presented.

As an editor of this book, I would like to thank all the authors for their contribution through the up-to-date research of their high-quality work. Lastly, I would like to express my thanks and gratitude to the InTech team for their support during the preparation of this book.

Dr. Vasilios N. Stavrou
Hellenic Naval Academy
Piraeus, Greece

Section 1

Introduction and Theoretical Review

Chapter 1

An Overview on Quantum Cascade Lasers: Origins and Development

Raúl Pecharromán-Gallego

Additional information is available at the end of the chapter

http://dx.doi.org/10.5772/65003

Abstract

This chapter presents an introductory review on quantum cascade lasers (QCLs). An overview is prefaced, including a brief description of their beginnings and operating basics. Materials used, as well as growth methods, are also described. The possibility of developing GaN-based QCLs is also shown. Summarizing, the applications of these structures cover a broad range, including spectroscopy, free-space communication, as well as applications to near-space radar and chemical/biological detection. Furthermore, a number of state-of-the-art applications are described in different fields, and finally a brief assessment of the possibilities of volume production and the overall state of the art in QCLs research are elaborated.

Keywords: quantum cascade lasers, review, history, operation, fundamentals, materials, photoacoustic spectroscopy, sensors, trace-gas detection, plasma species, cavity ring-down spectroscopy

1. Introduction

Quantum cascade lasers (QCLs) are based on a fundamentally different principle to 'classic' semiconductor lasers, that is, they use only one type of charge carriers, electrons, using intersubband transitions, so they can be called unipolar lasers. QCLs were conceived in the early 1970s. First, Esaki and Tsu [1] fabricated the first one-dimensional periodic potential multilayer by periodically varying the composition during epitaxial growth (superlattice). Later, QCLs were proposed by Kazarinov and Suris [2] and finally first demonstrated at Bell Laboratories in 1994 by Faist et al. [3]. Using superlattices leads to both quantum confinement and tunnelling phenomena, the basic processes in QCLs operation.

A conventional laser diode generates light by a single photon that is generated from an electron interband transition; this means that a high-energy electron in the conduction band recombines with a hole in the valence band, being the energy of the photon determined by the band-gap energy of the material system used. However, QCLs do not use bulk semiconductor materials in their optically active region, but a periodic series of thin layers of varying material composition forming a superlattice, which leads to an electric potential that changes across the length of the device (one-dimensional multiple quantum well confinement), splitting the band-permitted energies into a number of discrete electronic subbands, making electrons cascade down a series of identical energy steps built into the material during crystal growth, and emitting a photon at every step, unlike diode lasers, which emit only one photon over the equivalent cycle. With an appropriate design of the thickness of these layers, population inversion is achieved between discrete conduction band-excited states in the coupled quantum wells by the control of tunnelling, making laser emission possible. Therefore, the position of energy levels is mainly determined by the thickness of the layers, rather than the material, and thus allowing tuning the emission wavelength of QCLs over a wide range in the same material system. Thus, one electron emits a photon during every intersubband transition within the quantum well (QW) in the superlattice, and then can tunnel into the next period of the structure where another electron can be emitted, leading QCLs to outperform diode lasers operating at the same wavelength by a factor even greater than 1000 in terms of power.

Classically, a QCL is made of a periodic repetition of active sections, which consist of tunnel-coupled quantum wells and injector, where a miniband is formed. As **Figure 1** shows [4], from the injector miniband the electrons are injected into the upper laser energy level (4) of the active section, where the laser transition takes place. Afterwards, the lower laser energy level (3) is emptied by longitudinal optical emissions (LO emissions) and the electrons enter the next step by tunnelling.

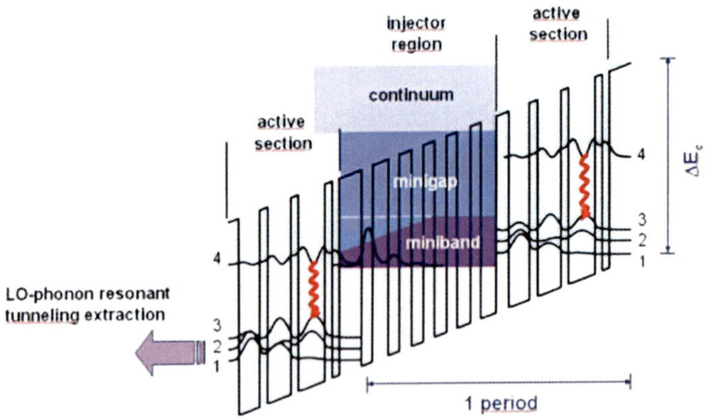

Figure 1. Typical conduction band structure of a QCL [4].

http://dx.doi.org/10.5772/65003

Combining materials in the active region, QCLs could be designed to emit at any wavelength over a wide range of the spectrum [5], as the emission wavelength is determined by quantum confinement. **Figure 2** shows a typical QCL in operation and some commercial examples. These structures are typically grown using either molecular beam epitaxy (MBE) or metal-organic chemical vapour deposition (MOCVD), being the most used growth mechanisms utilized to grow the alternated different semiconductor layers required for heterostructures fabrication on to a substrate. Ever since the first QCL was fabricated using InGaAs/InAlAs grown over InP substrate [1], other materials have been used in order to fabricate QCL structures, such as GaAs/AlGaAs, InGaAs/AlAsSb, InAs/AlSb, Si/SiGe and GaN-based materials, such as AlGaN/GaN and AlN/GaN.

Figure 2. Left: the invisible beam from a high-power quantum cascade laser lights a match in its path. The laser is 2.25 mm long and 17 μm wide, being placed in a cryostat at liquid-nitrogen temperature and emits an optical power in excess of 200 mW from each facet at a wavelength of 8.0 μm. Similar devices emit up to 600 mW at room temperature in pulsed mode. Right: front: mid-IR lasers (4–10 μm) for trace-gas analysis and IR molecular spectroscopy in the front; back: set-up examples.

First commercialized a decade later of their first demonstration [6], the key features of these lasers reside in the fact of their high optical power output and, on the other hand, their tuning range and room-temperature operation. Spectroscopy applications are related to gas detection and analysers (pollutants, components, etc.). Other practical uses include industrial control, plasma chemistry and detection, such as collision avoidance radar or poor visibility-driving condition aids. Finally, the 3 to 5μm atmospheric window would make QCLs perfect candidates for substitution of optical fibre in high-speed and free-space communications.

2. Fundamentals and operating principles

The main particularity in QCLs is the fact that instead of using bulk semiconductor materials in their optically active region, a periodic series of thin layers are used, that is, superlattices, consisting of a number of quantum well—barrier system equally spaced, introducing a multiple quantum well (MQW) leading to one-dimensional confinement allowing an electric potential variation (band splitting) that results in a number of discrete electronic subbands. In order to achieve the population inversion required for laser emission, it is necessary for a proper thickness and composition layer design. These confined energy levels depend on the layer thickness, so the tunability of the emission within the same material relies, in principle, on thickness variation, although multiwavelength QCLs emit by means of different materials within the same structure and multiple resonators [7, 8].

In classical semiconductor laser diodes, electrons and holes recombine across the band gap, thus, generating one photon per e^{-}-h^{+} pair recombined. However, in QCLs this is not the case, as an electron in the conduction band within a QW emits one photon whenever it undergoes an intersubband transition, that is, thermalization into lower-energy levels, emitting one photon. This electron could tunnel into the next period of the structure, where the mentioned transition happens again emitting another photon. This phenomenon of a single electron emitting multiple photons as it passes through different periods of the structure is called cascade, making the quantum efficiency of QCLs greater than unity and leading to their high optical output power.

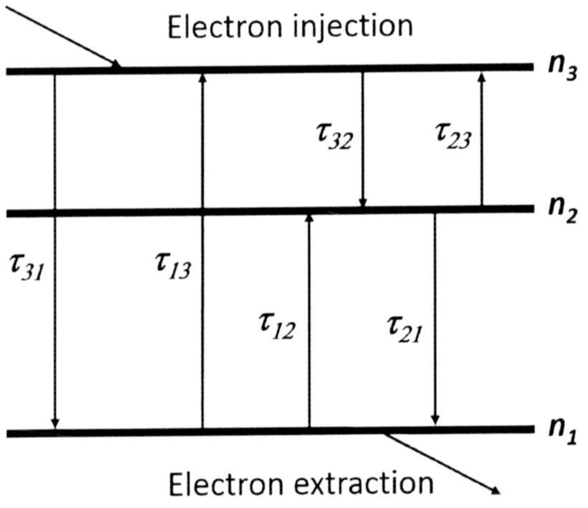

Figure 3. Three-level intersubband transition and scatterings considered in most QCLs.

QCLs are usually three-level lasers, whose level transitions are depicted in **Figure 3**. In these lasers, the wave function formation is faster than scattering between states; hence, the time-independent solutions of the Schrödinger equation can be applied, so the system can be modelled using rate equations. Each subband, i, is considered to have n_i electrons, and a scatter between the initial and the final subband, f, will have a scattering rate, W_{if}, and lifetime, τ_{if}. If no other subbands are populated, the rate equations are described as extraction and injection of electrons, respectively (depicted in **Figure 3**), whose values are both equal in steady-state conditions, where time derivatives are zero. Hence, the general rate equation for electrons in a subband i within an N level system is where $I = I_{ee} = I_{ie}$. At low temperature, absorption is near zero: hence, $\tau_{32} > \tau_{21}$, so $W_{21} > W_{32}$, and $n_3 > n_2$ leading to the existence of population inversion.

$$dn_1 / \mathrm{dt} = n_2 / \tau_{21} + n_3 / \tau_{31} - n_1 / \tau_{13} - n_1 / \tau_{12} - I_{\mathrm{ee}} \quad (1)$$

$$dn_2 / \mathrm{dt} = n_3 / \tau_{32} + n_1 / \tau_{12} - n_2 / \tau_{21} - n_1 / \tau_{23} \quad (2)$$

$$dn_3 / \mathrm{dt} = I_{\mathrm{ie}} + n_1 / \tau_{13} + n_2 / \tau_{23} - n_3 / \tau_{31} - n_3 / \tau_{23} \quad (3)$$

where I_{ee} and I_{ie} are the extraction and injection of electrons, respectively (depicted in **Figure 3**), whose values are both equal in steady-state conditions, where time derivatives are zero. Hence, the general rate equation for electrons in a subband i within an N level system is

$$dn_i / \mathrm{dt} = S_{f=1}^{N} n_j / \tau_{fi} - n_i S_{f=1}^{N} \tau_{if}^{-1} + \mathrm{I}(d_{i0} - d_{iN}) \quad (4)$$

where $I = I_{ee} = I_{ie}$. At low temperature, absorption is near zero:

$$n_3 / \tau_{32} = n_2 / \tau_{21} \quad (5)$$

hence, $\tau_{32} > \tau_{21}$, so $W_{21} > W_{32}$, and $n_3 > n_2$ leading to the existence of population inversion.

The scattering rate between two subbands strongly depends upon the overlap of the wave functions and energy spacing between the subbands. In order to decrease W_{32}, the overlap of the upper and lower laser levels is reduced. This is often achieved through designing the layer thickness such that the upper laser level is mostly localized in the left part of the well of the 3QWs active region, while the lower laser level wave function is made to mostly reside in the central and right part of the wells, leading to the so-called diagonal transition. A vertical transition is one which the upper laser level is rather localized in the central and right part of the wells, increasing the overlap and, therefore, W_{32}, reducing the population inversion but increasing the strength of the radiative transition and hence the gain. On the other hand, in order to increase W_{21}, the lower laser level and the ground level wave functions are designed to obtain a good overlap, and the energy spacing between the subbands is designed such that it is equal to the longitudinal optical (LO) phonon energy so that the resonant LO phonon-electron scattering can depopulate the lower laser level. For instance, the LO-phonon energy is around 36 meV in GaAs (which is comparable to the room temperature $k_B T$ value of around 26 meV) and around 91 meV in GaN [9].

Tunable semiconductor lasers can be produced by using multiple resonators or multisection injection devices [10–12]. In fact, tunable QCLs have been demonstrated [7] some time ago. Moreover, a QCL with a heterogeneous cascade containing two substacks previously optimized to emit at 5.2 and 8.0 μm, respectively, was presented by Gmachl et al. [13]. On the other hand, single-mode tunable QC distributed feedback lasers emitting between 4.6 and 4.7μm wavelength have been reported [14]. These lasers were pulsed, continuously tunable single-mode emission and were achieved from 90 to 300 K with a tuning range of 65 nm and a peak output power of approximately 100 mW at room temperature—so the lasers described by Pellandini et al. [12], Gmachl et al. [13] and Köhler et al. [14] were laser sources for the mid-infrared region. In order to realize near-infrared QCLs, optical non-linearity in intersubband lasers has been used to design such lasers emitting at 4.76 μm, with third harmonic and second harmonic generation at 1.59 and 2.38 μm, respectively [15].

3. Materials used and growth methods

The first QCL in 1994 used InAlAs as the cladding layers due to its low refractive index of around 3.20. The core region, which includes active and injector regions, usually has 500 stacking layers consisting of alternative InGaAs and InAlAs layers with total thickness about 1.5–2.5 μm. The average refractive index in this region can be calculated according to the volume fraction of these two constituent materials and is often around 3.35, which is clearly higher than the cladding layers [16]. The confinement factor with typical Np of approximately 30 is usually around 0.5. To reduce the optical loss, the cladding layers are usually doped with a low concentration of 5×10^{16} cm^{-3} and the separate confinement heterostructures (SCHs) are implemented with InGaAs of high refractive index to increase the optical confinement factors. In later designs, the InAlAs cladding layers were replaced by InP because it has a lower refractive index of around 3.10 and a higher thermal conductivity for better heat treatment, which is a critical step for the device performance [16].

So far, the emission wavelength of QCLs has been extended from the near infrared (around 100 THz) to terahertz regimes. While the longest demonstrated wavelength is 1.6–1.8 THz with $GaAs/Al_{0.1}Ga_{0.9}As$ system at 80 K under continuous-wave (CW) operation [17], the shortest wavelength has been extended to 3 μm with $In_{0.53}Ga_{0.47}As/AlAs_{0.56}Sb_{0.44}$ system at 300 K under pulsed operation [18]. The highest output power of short-wavelength QCL (4.6 μm) under CW operation at room temperature has been demonstrated (100 mW and maximum temperature 105°C). All of these state-of-the-art QCLs have been grown by MBE so far, but for industry MOCVD is preferable for mass production. As a result, attempts have been made to grow QCLs by MOCVD. Until now, there are only three groups successfully reporting the growth of QCLs by MOCVD: an InP-based QCL ($\lambda \approx 8.5$ μm) operating in pulsed mode at room temperature, with low-temperature threshold current density in the region of 1500 A/cm^2 [19], an $In_{0.53}Ga_{0.47}As/In_{52}Al_{0.48}As$ QCL ($\lambda \approx 8.5$) operated in continuous-wave operation at room temperature with an output power of 5.3 mW [20], and Diehl et al. [21] managed to grow a QCL working in continuous-wave mode above 370 K, with an optical output power of 312 mW at room temperature and an emission wavelength of 5.29 mm. More effort is still required for MOCVD growers to develop more advanced techniques to compete with MBE. Having said that, room-temperature operation in InAs/AlSb QCLs has been achieved at 4.5 μm [22]; more recently, MBE-grown InAs/AlSb on n-InAs (100) substrate QCL operating at 8.9 μm has been reported, with a maximum operating temperature of 305 K [23].

QCLs provide one possible method of realizing high-efficiency light emitters in indirect band gap materials such as silicon. Electroluminescence at terahertz frequencies from Si/SiGe intrasubband transitions has been demonstrated [24], and recently silicide low-loss (down to 2 cm^{-1}) waveguides were designed [25, 26]. **Figure 4** shows a transmission electron microscopy (TEM) image from a Si/SiGe QCL.

Figure 4. TEM image from a Si/SiGe quantum cascade structure, consisting of 600 periods of 6.5 nm $Si_{0.7}Ge_{0.3}$ quantum wells with 2-nm strained-Si barriers, all grown on top of a $Si_{0.8}Ge_{0.2}$ virtual substrate. The quantum cascade laser has a total thickness of 5 mm.

Although GaN-based materials have not been employed to fabricate QCLs, they are also promising materials to be used as such devices. III–V nitrides are known in their wurtzite structure to possess a large spontaneous polarization and piezoelectric constants. As a result, two-dimensional (2D) charges build up at nitride heterointerfaces where the polarization discontinuities occur, causing strong built-in electric fields [27]. Furthermore, GaN is a material with large LO-phonon energy, leading to a thermal population reduction of the lower laser state, a feature desirable for high-temperature operation of terahertz QCLs, as proposed by Diehl et al. [21]. On the other hand, ultrafast LO-phonon scattering in GAN/AlGaN QWs can be useful in order to rapidly depopulate the lower laser state [28, 29]. Lastly, the large LO-phonon energy can also increase the lifetime of the upper laser state by reducing the relaxation of electrons with higher in-plane kinetic energy via emission of a LO phonon. Using low-pressure MOCVD, GaN/AlGaN active regions for QCLs have been grown by Huang et al. [30]. GaN/AlGaN active layers are depicted in **Figure 5**.

Finally, typical phonon frequency redshift is a key indicator of good periodicity of MOVPE-grown GaN/AlGaN QCL structures, 822 cm^{-1} in the superlattice has been measured, indicating a redshift with respect to the single AlGaN layer [31].

AlN/GaN compound is another possible material to be used to fabricate quantum cascade structures. A suitable method of fabrication is hot-wall epitaxy (HWE) [32], a low-cost, convenient and scalable technique, where the epitaxial layers are grown under conditions as near as possible to thermodynamic equilibrium, allowing a minimum material loss. Inoue et al. [33] grew short-period superlattices consisting of five periods of GaN wells of 10 (or nine) molecular layers (MLs) each with 1 ML AlN barriers, which was designed to emit photons at a wavelength within the mid-infrared range (around 5 μm) [34, 35].

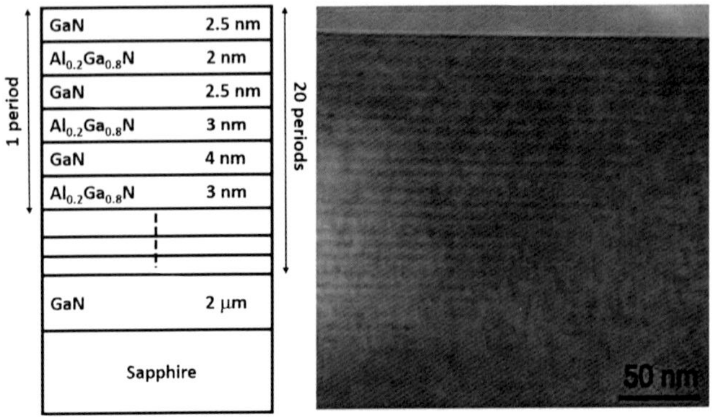

Figure 5. Schematic diagram of the AlGaN QCL active layer structure (left) and cross-sectional TEM (right) of the 20-period GaN/AlGaN MQW MQL structure.

4. Applications

We could say that QCLs, in short, are eligible to those applications where a powerful and reliable mid-infrared source is required. For instance, most chemical compounds have their fundamental modes in the mid-infrared region (3–15 μm) of the electromagnetic spectrum, thus making this range of paramount importance for gas sensing and spectroscopy applications. The so-called two 'atmospheric windows' are two windows corresponding to the ranges 3–5 and 8–12 μm, at which the atmosphere happens to show a high transparency leading to remote sensing and detection in those windows. In this section, a brief description of the current applications of QCLs is elaborated, being summarized as subsequently.

4.1. Trace-gas detection by optical methods in the mid-infrared

Most trace gases of importance, from products of fossil fuel burning to constituents of human breath, have telltale absorption features in this wavelength range, that is, their 'fingerprint' region of the spectrum, as a result of molecular rotational-vibrational transitions [16]. Narrow-linewidth, tunable semiconductor lasers in this wavelength range are used to spectrally map out and qualitatively and quantitatively detect these trace gases, by a measurement technique called tunable infrared laser diode absorption spectroscopy (TILDAS) [36]. A schematic representation of a TILDAS is shown in **Figure 6**. The advantage of TILDAS is its high sensitivity and specificity, usually combined with the advantages of the solid-state device approach.

This technique has been successfully introduced in distributed feedback quantum cascade laser (DFB-QCL) structures [37]. DFB-QCLs were first introduced in 1997 [38], providing continuously tunable single-mode laser output, and were demonstrated for the first time in trace-gas sensing applications 1 year later [37]. DFB-QCLs operate as follows: a grating with period L is incorporated into the wavewide, lowering the threshold gain (by reducing the outcoming loss) for a different wavelength close to the Bragg wavelength:

$$\lambda = 2 \cdot n_{\text{eff}}(T) \cdot L \tag{6}$$

where n_{eff} is the effective refractive index of the waveguide mode, and is a function of temperature. As temperature increases, wavelength shifts to longer values. Changing the heat-sink temperature could control laser temperature; however, the process is slow due to the fact that it implies adiabatically temperature change of a large volume. Tuning rates increase with heat-sink temperature current from 0.3 to 0.4 nm K^{-1}, for a laser emitting at approximately 5.2 μm, and from 0.4 to 0.65 nm K^{-1}, for a laser whose emitting wavelength is around 8 μm [16].

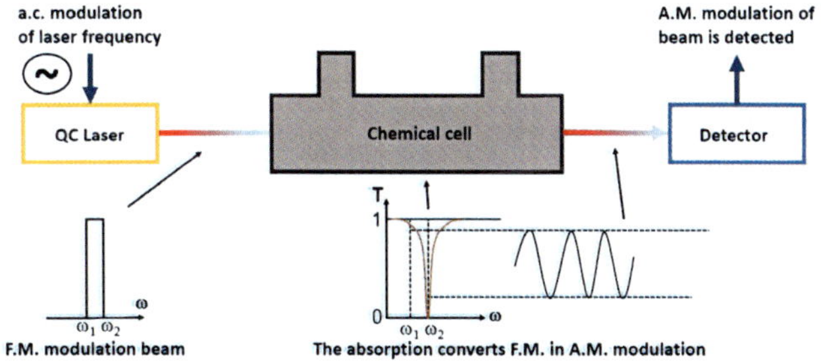

Figure 6. Depiction of a tunable infrared laser diode absorption spectroscopy (TILDAS).

4.2. Absorption measurements

In direct absorption measurements, the change in intensity of a beam is recorded as the latter crosses a sampling cell where the chemical to be detected is contained, making this measurement technique quite simple. In a version of this technique, the light interacts with the substance through the evanescent field of a waveguide or an optical fibre.

Namjou et al. [37] used the first DFB-QCL for gas detection purposes. The laser was operated at room-temperature conditions in order to measure mid-infrared (1 around 7.8 μm) absorption spectra of gases N_2O and CH_4, diluted in N_2 and prepared in a 10-cm long single pass gas cell, using a wavelength-modulation technique. The noise equivalent sensitivity limit of the measurement was 50 ppm. Sharpe et al. [39] used a DFB-QCL emitting at 5.2 and 8.5 μm to carry out direct absorption measurements of NO and NH_3, respectively. Williams et al. [40] measured the intrinsic linewidth of several DFB-QCLs emitting at around 8 μm, by observing fluctuation of the collected optical intensity when the laser beam was passing through a sample cell containing N_2O, with a well-known absorption features as the laser was being tuned to the side of one such absorption line. Finally, Kosterev et al. [41] developed a variable-duty cycle and quasi-CW frequency-scanning technique for DFB-QCLs, which relieves many of the thermally induced effects of pulsed operation. Combining this laser with a 100-m multipass gas cell and zero-air background subtraction, a detection sensitivity close to 1 ppb concentration levels was achieved; these authors also demonstrated, by using QC-DFB lasers, high-sensitivity detection of simple molecules when a spectral resolution of approximately 300 MHz is sufficient [42].

4.3. Cavity ring-down spectroscopy

Cavity ring-down spectroscopy (CRDS) is used to measure the concentration of some light-absorbing substance, typically a gas. A short pulse of laser light is injected into the cavity, bouncing back and forth between the mirrors facing each other that make up the cavity. Typically 0.1% of the laser light comes out of the cavity and can be measured whenever the light hits one of the mirrors. Hence, as some light is lost on each reflection, the amount of light hitting the mirror decreases each time, leaking a percentage through. An absorbent medium is placed in the cavity, making light undergo fewer detections before it is extinguished. In short, CRDS measures the time it takes for the light to drop for a certain percentage of its original amount, and this ring-down time can be converted to a concentration, having two main advantages: fluctuations in the laser do not affect the measurement (the ring-down time does not depend upon the brightness of the laser), and due to its long pathlength, it is very sensitive, since the light reflects many times between the mirrors.

Paldus et al. [43] reported the use of a 126 mW CW operated DFB-QCL emitting at around 8.5 μm for CRDS of ammonia diluted in nitrogen. A sensitivity of 3.4×10^{-9} cm^{-1} Hz-1/2 was achieved for ammonia in nitrogen at standard temperature and pressure, which corresponds to a detection limit of 0.25 ppbv.

4.4. Photoacoustic spectroscopy

The photoacoustic effect was discovered in 1880 by Alexander Graham Bell, who showed that thin discs emit sound when exposed to a beam of sunlight that was quickly interrupted with a rotating slotted disc. The absorbed energy from the light was transformed into kinetic energy in the sample and so a local heating and a pressure wave of sound arose. Later, Bell demonstrated the fact that materials exposed to different regions of the solar spectrum, that is, infrared and ultraviolet, can produce sounds too. Hence, by measuring the sound at different wavelengths, a photoacoustic spectrum from a sample can be recorded, so that it can be used to identify the absorbing components of the sample. This technique can be used to study solids, liquids and gases.

One of the uses of photoacoustic spectroscopy is the study of gas concentrations at the parts per billion or even parts per trillion levels. Although most photoacoustic detectors do not differ much from the original Bell's set-up, some enhancements have been incorporated in order to increase the sensitivity, such as enclosing the gaseous sample in a cylindrical chamber, and amplifying the sound signal by tuning the modulation frequency to an acoustic resonance of

http://dx.doi.org/10.5772/65003

the sample cell, by using high-sensitive microphones together with lock-in techniques and utilizing intense lasers instead of sunlight to illuminate the sample.

Paldus et al. [44] reported photoacoustic spectroscopy on NH_3 and H_2O diluted in N_2 using a CW DFB-QCL emitting at a wavelength of 8.5 μm. The noise-limited minimum detectable concentration of NH_3 was 100 ppbv for an integration time of 1 s. Lastly, Nägele et al. [45] built a multicomponent (ethane, methanol and ethanol) trace-gas monitoring system using QCLs as pump sources and a multipass photoacoustic cell.

4.5. Other spectroscopies

Other types of spectroscopies have also been used in the context of QCLs detection.

First, lamb-dip spectroscopy is a useful technique to study the spectra from polyatomic molecules. When a monochromatic light with a given wavelength passes through a chemical cell in a set-up similar to that depicted in **Figure 6**, a Gaussian absorption spectrum centred in the wavelength comes up. If we measure the absorption of a light beam (probe beam) by passing through two beams (pump and probe beams) of the same wavelength from opposite directions, by using a beam splitter, a less intense absorption is observed at that wavelength. This reduction in intensity appears as a dip, the so-called lamb-dip, and its position gives the location of the transition wavelength having no Doppler shift (Doppler-free in the absorption curve). Samman et al. [46] used a CW-operated DFB-QCL emitting at around 5.2 mm to obtain sub-Doppler resolution-limited saturation features in a lamb-dip experiment on NO. These lamb dips appeared as transition spikes with full-width at half maximum (FWHM) values around 4.3 MHz.

Gittins et al. [47] used a multimode Fabry-Perot-type QCL emitting at around 8.0 μm wavelength for quantitative backscatter absorption measurements on isopropanol vapour. They developed and employed a pseudorandom code modulation of the laser, and explored its use for different absorption LIDAR (laser infrared detection and ranging).

Finally, Sonnenfroh et al. [48] used DFB-QCLs (1 around 5.4 μm) in quasi-CW mode close to room temperature in conjunction with a balanced radiometric detection technique to achieve high sensitivity, whereas Hvozdara et al. [49] demonstrated the first use of a GaAs-based QCL for gas-sensing applications.

4.6. Direct detection applications: drug, explosive, plasma species and aerospace and military

Drug detection is another important application field of QCL-based detection systems [50]. Lu et al. [51] demonstrated, by integrating an optoelectronic terahertz microsource into a glass-substrated microchip within the near-field distance, a compact, label-free and non-invasive microbiosensing terahertz device that allowed detection of illicit drug powders with weight in the order of the nanogram.

The detection of explosive fingerprints is of great importance for security reasons. The detection of such substances has a number of drawbacks as explosive molecules are heavy, complex and have thousands of rotational and vibrational transitions. Furthermore, broadband features with no separation observed at low pressures happen in measured transmission spectra. Different explosives have, therefore, unique terahertz spectral fingerprints. Williams et al. [52] developed a high-power QCL emitting at around 4.4 THz. These lasers are based on a resonant-phonon depopulation scheme, and use a semi-insulating surface-plasmon waveguide. They managed a maximum power of ~248 and 138 mW pulsed and CW, respectively. van Neste et al. [53] have used two QCLs operating simultaneously with tunable wavelength windows matching the absorption peaks of analytes in order to improve sensitivities of standoff spectroscopy techniques, leading to a sensitivity of 100 ng/cm^2 and a standoff detection distance of 20 m for surface-adsorbed analytes such as explosives and tributyl phosphate.

Molecular plasmas are tools in plasma-enhanced chemical vapour deposition and in etching systems to deposit or remove thin films. The analysis of the precursor gas fragmentation and the monitoring of plasma reaction products is the key to an improved understanding and control of these chemical-active discharges. This observation can be done by using absorption spectroscopy methods in the mid-infrared spectral region. Recently, a QCL-based absorption spectroscopic system, the quantum cascade laser measurement and control system (Q-MACS), was developed and used to study industrial process plasmas and for environmental studies, and its applicability for monitoring online processes has been proven [54].

Finally, QCLs are also applied to aerospace and military uses. For instance, the first of a new generation of miniature tunable laser mid-IR spectrometers operating at room temperature for in situ measurement of atmospheric and evolved planetary gases started to develop not too long ago. These devices are based on newly available room-temperature QCL sources in the 3–12 μm wavelength region and have immediate applications to Mars, Titan, Venus and Europa missions, being operated on a descending or penetrating probe, lander, rover or aerobot, and would consume only a few watts of power, with a weight less than 1 kg [55]. Furthermore, features of QCL make them good laser sources to carry out non-destructive-imaging engine combustors, where failure mechanisms, engine performance in aircraft and moisture content in jet fuel are examined. A low thermal noise background is required for such purpose, and ceramic ports exist already on such platforms transparent to terahertz radiation. Scale model radar cross-section experiments have also been carried out in the aerospace industry, to replace the bulky and extremely inefficient molecular lasers with QCLs, at the University of Massachusetts Lowell Sub-Millimetre Wave Laboratory [56], to perform scale model radar cross-section measurements for phenomenology and target-recognition database formation.

The infrared spectrum is also used in the aerospace industry for detection purposes, for instance, in the infrared scene generation, which is presently a critical technology for testing of infrared-imaging systems, for example, in infrared-guided missile systems, and QCLs could have an application in this technology to replace large- and slow-response 'resistor banks' with tailored design QCLs, to mimic the thermal background of a given scene [57]. Moreover, coherent transceiver using a terahertz quantum cascade laser (TQCL) as the transmitter and an optically pumped molecular laser as the local oscillator has been used for imaging purposes along with inverse synthetic aperture radar (ISAR), in which the range of the target was limited by the TQCL power (around 10–4 W) and indoor atmospheric attenuation at 2.408 THz, leading to a coherence length of the transmitter of up to several hundred metres [58].

5. State of the art and outlook

QCLs are also currently facing a number of challenges, which will be summarized below.

First, the extensions of the wavelength range into the far-infrared. Rochat et al. [59] grew far-infrared QC structures based on a vertical transition active region emitting at l around 76 μm. This is a challenging issue, so long as population inversion is difficult to attain for such long wavelengths. Energy levels are generally so close that selective injection into a single level is difficult. Furthermore, LO-phonon scattering is replaced by electron-electron scattering as the dominant electron relaxation mechanism, which is more difficult to model accurately. Moreover, wavewide losses are expected to be high, although even very low-temperature QCL operation would be a great accomplishment due to the lack of narrow band and high-power compact sources in the far-infrared wavelength range. In fact, temperature is a key player when it is required to achieve a certain emission wavelength as **Figure 7** shows.

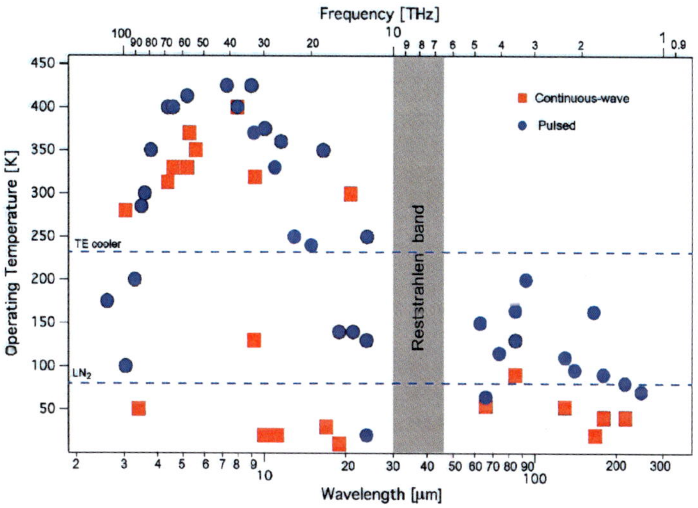

Figure 7. Operating temperature plot as a function of the emission wavelength/frequency for QCLs [60].

Another challenge would also be to fibre-optic wavelengths. In fact, the fabrication of large optical waveguides, powers of 14 and 5 W at respective temperatures of 15 and 280 K, is demonstrated at a wavelength of approximately 5.2 μm [61].

On the other side of the IR spectrum, reaching shorter wavelengths is another motivation for a more selection of materials. Work on this respect has also be done on GaN-based devices [61], with Gmachl et al. [62] having measured intersubband optical absorption in narrow (15–30 Å wide), GaN/AlGaN quantum wells grown by MBE on sapphire substrate. Peak absorption wavelengths ranged from 4.2 μm for 30Å wide wells to 1.77 μm for a 15Å wide well. On the other hand, Iizuka et al. [63] reported ultrafast intersubband relaxation (less than 150 fs) at a wavelength of 4.5 μm in $Al_{0.65}Ga_{0.35}N/GaN$ MQWs, with as many as 200 QWs, making this result promising for fabrication ultrafast optical switches.

Doping level of the active region is a key optimization parameter that determines maximum drive current, optical losses and threshold current. Faist et al. [61] presented a systematic change of the active region doping in an InAlAs-InGaAs/InP lasers emitting at 9 μm. On the other hand, the wavelength tunability of each QC-DFB laser is especially limited to cover the entire molecular absorption spectrum of volatile organic compounds and hydrocarbons, which could be addressed by separating the gain medium from the wavelength-selective element [61].

Other challenges involve the low conversion efficiencies (<1%) between electrical and optical power, also known as wall plug efficiency (WPE) [64], and also the unavoidable fact that far-infrared (terahertz) QCLs lack proper performance at room-temperature operation. Belkin et al. [65] have recently reviewed recent research that has led to a new class of QCL light sources that has overcome these limitations leading to room-temperature operation in the terahertz spectral range, with nearly 2 mW of optical power and significant tunability by using intra-cavity THz difference-frequency generation (DFG) in dual-wavelength mid-IR QCLs. However, Lu et al. [66] recently presented a strong-coupled strain-balanced quantum cascade laser design for efficient THz generation based on intra-cavity difference-frequency generation, demonstrating continuous-wave, single-mode THz emissions with a wide-frequency tuning range of 2.06–4.35 THz and an output power up to 4.2 μW at room temperature from two monolithic three-section sampled grating distributed feedback-distributed Bragg reflector lasers.

Furthermore, Burghoff et al. [67] demonstrated frequency combs based on terahertz QCLs, combining high power of lasers with the broadband capabilities of pulsed sources. By fully exploiting the quantum-mechanically broadened gain spectrum available to these lasers, 5 mW of terahertz power spread across 70 laser lines can be generated. Therefore, the radiation is sufficiently powerful to be detected by Schottky-diode mixers, and will lead to compact terahertz spectrometers.

About gas detection, Harrer et al. [68] combined the operation mode and low-divergence emission from QCLs with two-dimensional array integration with multiple emission and detection frequencies leading to detecting propane concentrations of 0–70 and 0–90% for isobutane at a laser operation wavelength of 6.5 μm using a 10 cm gas cell in double-pass configuration.

Finally, possibly the best way to assess the current state of the art for volume production of QCLs is to start from the market requirement from the end-user point of view. An example is the continuous emission monitoring (CEM) market, which includes engine emission and power-plant stack monitoring, is fiercely competitive, with a range of different gas-sensing technologies, such as Fourier transform spectrometry, chemiluminescence, non-dispersive infrared, and so on, each trying to increase its market share. Among these technologies, QCL-based gas sensor technology is still in its infancy, since although predictions indicate a great improvement in performance, much demonstration and convincing is still required in this conservative market. Moreover, it is difficult to justify a system whose price is above the market, despite better performance, due to the risk of an unproven and new technology. Hence, a QCL device might be priced below the market, which can have negative consequences that can reach the component supplier. Customers are usually reluctant to acquire a new sensor having other well-proven systems already in the market. Therefore, demonstration of the unique advantages of the QCL-based devices is often necessary, including specifications, performance, mean time to failure (MTTF) parameters, traceability and warranty period with a significant impact [69]. Implementation of high-volume production and proper validation processes on QCLs are producing positive outcomes. Along with a noticeable improvement in gas-sensing performance parameters, volume-cost reduction is allowing QCL systems price decrease to meet market demands. By consistently tackling the challenges of the QCL volume manufacturing, carrying out severe testing procedures, implementing quality-control systems and reaching adequate device costing, QCL developers and manufacturers can complete and come into the marketplace.

Finally, continuing to build consumer confidence in QCL commercial products, it should be easier for customers to digest the unavoidable contract 'warranty' and 'product liability' clauses found in supply contracts [69].

6. Summary

In this chapter, an overview of quantum cascade lasers (QCLs) is presented. A historical introduction is first introduced. The basic features of QCLs are outlined, as well as a brief description of the issues that this work deals with. The operation and fundamentals of QCLs are also described. An analysis of the operation of these structures is included. Basically, the use of superlattices and tunability due to the layer thickness are the key features, in conjunction with the intersubband tunnelling transition – cascade. The QCLs are usually three-level lasers. Rate equations are also included. Furthermore, introducing a graphical brief timeline (**Figure 8**), including some of the most important milestones achieved in the world of QCL, would be helpful in this section.

An overview over the materials used for these devices is also included, that is, InGaAs and InAlAs layers, InAs/AlSb, Si/SiGe and GaN-based.

QCLs have a wide range of applications: first, trace-gas detection by optical methods in the mid-infrared, the great suitability of the TILDAS techniques and DFB-QCLs in trace-gas-sensing applications.

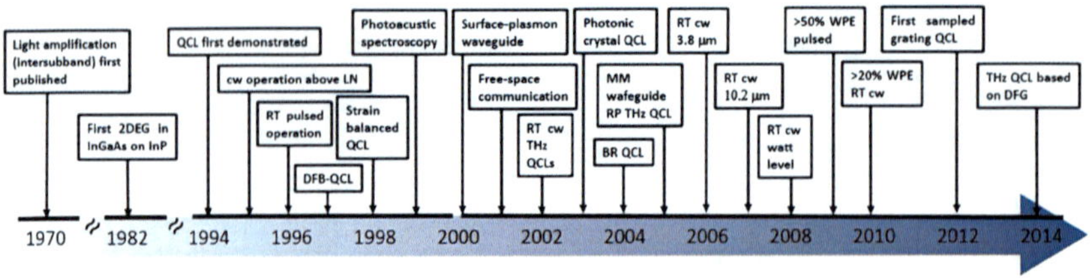

Figure 8. Achievements of QCLs over time.

Cavity ring-down spectroscopy is another technique used for gas detection and briefly described above. QCLs are also applied in photoacoustic spectroscopy, to study gas concentrations at the parts per billion or even parts per trillion levels. Applications in gas detection using QCLs with other types of spectroscopy and variations, such as lamb-dip spectroscopy, radiometric detection techniques, and so on, are mentioned.

The latest advances in QCL applications are also described. The extension of the wavelength range into the far-infrared but also shorter wavelengths, fibre-optic wavelengths and fabrication of large optical waveguides, doping issues, room-temperature operation problems, frequency combs based on terahertz devices, new insights on gas detection and finally current state of the art for volume production of QCLs are mentioned.

Author details

Raúl Pecharromán-Gallego*

Address all correspondence to: anubis_rpg@hotmail.com

The Last Push Consulting, Heverlee, Leuven, Belgium

References

[1] Esaki L, Tsu R. Superlattice and negative differential conductivity in semiconductors. IBM Journal of Research and Development. 1970;14(1):61–65. DOI: 10.1147/rd.141.0061.

[2] Kazarinov R, Suris R. Possibility of the amplification of electromagnetic waves in a semiconductor with a superlattice. Soviet Physics—Semiconductors. 1971;5(4):707–709.

[3] Faist J, Capasso F, Sivco D L, Sirtori C, Hutchinson A L, Cho A Y. Quantum cascade laser. Science. 1994;264:553–556. DOI: 10.1126/science.264.5158.553.

[4] Markus-Christian Amann. Semiconductor Technology (E26) [Internet]. [Updated: 2016]. Available from: https://www.wsi.tum.de/Research/AmanngroupE26/AreasofResearch/tabid/115/Default.aspx [Accessed: 12-05-2016].

[5] Capasso F, Gmachl C, Sivco D L, Cho Q Y. Quantum cascade lasers. Physics World. 1999;12(6):27–33. DOI: http://dx.doi.org/10.1088/2058-7058/12/6/26.

[6] Alpes Lasers SA. Alpes Lasers—Quantum Cascade Lasers [Internet]. [Updated: 2016]. Available from: http://www.alpeslasers.ch/ [Accessed: 12-05-2016].

[7] Tredicucci A, Gmachl C, Capasso F, Sivco D L, Hutchinson A L, Cho A Y. A multiwavelength semiconductor laser. Nature. 1998;396:350–353. DOI: 10.1038/24585.

[8] Kosterev A A, Tittel F K, Köhler R, Gmachl C, Capasso F, Sivco D L, Cho A Y, Wehe S, Allen M G. Thermoelectrically cooled quantum-cascade-laser-based sensor for the continuous monitoring of ambient atmospheric carbon monoxide. Applied Optics. 2002;41(6):1169–1173. DOI: 10.1364/AO.41.001169.

[9] Edgar J H, Strite S, Akasaki I, Amano H, Wetzel C, editors. Properties processing and applications of gallium nitride and related semiconductors. 1st ed. UK: EMIS datareviews series no 23. INSPEC; 1999. 830 p. DOI: ISBN-10: 0852969538. ISBN-13: 978-0852969533.

[10] Beernik K J, Thornton R L, Chung H F. Low threshold current dual wavelength planar buried heterostructure lasers with close spatial and large spectral separation. Applied Physics Letters. 1994;64(9):1082–1084. DOI: 10.1063/1.110939.

[11] Dutta N K, Cella T, Zilko J L, Ackerman D A, Piccirilli A B, Greene L I. InGaAsP closely spaced dual wavelength laser. Applied Physics Letters. 1986;48(25):1725–1726. DOI: 10.1063/1.96815.

[12] Pellandini P, Stanley R P, Houdre R, Oesterle U, Ilegems M, Weisbuch C. Dual-wavelength laser emission from a coupled semiconductor microcavity. Applied Physics Letters. 1997;71(7):864–866. DOI: 10.1063/1.119671.

[13] Gmachl C, Sivco D L, Baillargeon J N, Hutchinson A L, Capasso F, Cho A Y. Quantum cascade lasers with a heterogeneous cascade: two-wavelength operation. Applied Physics Letters. 2001;79(5):572–574. DOI: 10.1063/1.1383806.

[14] #Köhler R, Gmachl C, Tredicucci A, Capasso F, Sivco D L, Chu S N G, Cho A Y. Single-mode tunable, pulsed, and continuous wave quantum-cascade distributed feedback lasers at $\lambda \cong 4.6$–4.7 μm. Applied Physics Letters. 2000;79(9):1092–1094. DOI: 10.1063/1.125987.

[15] Banerjee S, Spencer P S, Shore K A. Design of a tunable quantum cascade laser with enhanced optical non-linearities. IEE Proceedings Optoelectronics. 2006;153(1):40–42. DOI: 10.1049/ip-opt:20050061.

[16] Gmachl C, Capasso F, Sivco D L, Cho A Y. Recent progress in quantum cascade lasers and applications. Report on Progress in Physics. 2001;64(11):1533–1601. DOI: 10.1088/0034-4885/64/11/204.

[17] Walther C, Scalari G, Faist J, Beere H, Ritchie D. Low frequency terahertz quantum cascade laser operating from 1.6 to 1.8THz. Applied Physics Letters. 2006;89(23): 231121–3. DOI: 10.1063/1.2404598.

[18] Revin D G, Cockburn J W, Steer M J, Airey R J, Hopkinson M, Krysa A B, Wilson R, Menzel S. InGaAs/AlAsSb/InP quantum cascade lasers operating at wavelengths close to 3 μm. Applied Physics Letters. 2007;90(2):021108–3. DOI: 10.1063/1.2431035.

[19] Green R, Roberts J, Krysa A, Wilson L, Cockburn J, Revin D, Zibik E, Carder D, Airey P. MOVPE grown quantum cascade lasers. Physica E: Low-dimensional Systems and Nanostructures. 2004;21(2-4):863–866. DOI: 10.1016/j.physe.2003.11.133.

[20] Liu Z., Wasserman D, Howard S S, Hoffman A J, Gmachl C F, Wang X, Tanbun-Ek T, Cheng L, Choa F-S. Room-temperature continuous-wave quantum cascade lasers grown by MOCVD without lateral regrowth. IEEE Photonics Technology Letters. 2006;18(12):1347–1349. DOI: 10.1109/LPT.2006.877006.

[21] Diehl L, Bour D, Corzine S, Zhu J, Höfler G, Lončar M, Troccoli M, Capasso F. High-temperature continuous wave operation of strain-balanced quantum cascade lasers grown by metal organic vapor-phase epitaxy. Applied Physics Letters. 2006;89(8): 081101–3. DOI: 10.1063/1.2337284.

[22] Teissier R, Barate D, Vicet A, Alibert C, Baranov B A, Marcadet X, Renard X, Garcia M, Sirtori C, Revin D, Cockburn J. Room temperature operation of InAs/AlSb. Applied Physics Letter. 2004;85(2):167–169. DOI: 10.1063/1.1768306.

[23] Ohtani K, Fujita K, Ohno H. Room-temperature InAs/AlSb quantum-cascade laser operating at 8.9 μm. Electronics Letters. 2007;43(9):520–522. DOI: 10.1049/el:20070251.

[24] Lynch S A, Bates R, Paul D J, Norris D J, Cullis A G, Ikonic Z, Kelsall R W, Harrison P, Arnone D D, Pidgeon C R. Intersubband electroluminescence from Si/SiGe cascade emitters at terahertz frequencies. Applied Physics Letters. 2002;81(9):1543–1545. DOI: 10.1063/1.1501759.

[25] De Rossi A, Carras M, Paul D J. Low-loss surface-mode waveguides for terahertz Si–SiGe quantum cascade lasers. IEEE Journal of Quantum Electronics. 2006;42(2):1233–1238. DOI: 10.1109/JQE.2006.883496.

[26] Paul D J. Si/SiGe heterostructures: from material and physics to devices and circuits. Semiconductor Science and Technology. 2004;19(10):75–108. DOI: 10.1088/0268-1242/19/10/R02.

[27] Pecharromán-Gallego R. Investigations of the luminescence of GaN and InGaN/GaN quantum wells [thesis]. UK: University of Strathclyde; 2004. 216 p. Available from: http://suprimo.lib.strath.ac.uk/primo_library/libweb/action/display.do?tabs=detailsTab&ct=display&fn=search&doc=SUVOY682303&indx=1&recIds=SUVOY682303&recIdxs=0&elementId=0&renderMode=poppedOut&displayMode=full&frbrVersion=&dscnt=1&scp.scps=scope%3A%28SU%29%2Cscope%3A%28cla%29&frbg=&tab=local&dstmp=1463141375653&srt=rank&mode=Ba-

sic&dum=true&tb=t&vl(freeText0)=Pecharroman&vid=SUVU01 DOI: Thesis no : T10902.

[28] Sun G, Soref R A, Khurgin J B. Active region design of a terahertz GaN/Al0.15Ga0.85N quantum cascade laser. Superlattices and Microstructures. 2005;37(2):107–113. DOI: 10.1016/j.spmi.2004.09.046.

[29] Heber J D, Gmachl C, Ng H M, Cho A Y. Comparative study of ultrafast intersubband electron scattering times at ~1.55 μm. Applied Physics Letters. 2002;81(7):1237–1239. DOI: 10.1063/1.1500412.

[30] Huang G S, Lu T C, Yao H H, Kuo H C, Wang S C, Sun G, Lin C-W, Chang L, Soref R A. GaN/AlGaN active regions for terahertz quantum cascade lasers grown by low-pressure metal organic vapor deposition. Journal of Crystal Growth. 2007;298(1):687–390. DOI: 10.1016/j.jcrysgro.2006.10.106.

[31] Wang S C, Soref R, Sun G. Fabrication and characteristics of GaN/AlGaN multilayer structure for terahertz quantum-cascade laser. SPIE Proceedings. Terahertz Physics, Devices, and Systems. 2006;6373:637309. DOI: 10.1117/12.685678.

[32] Lopez-Otero A. Hot wall epitaxy. Thin Solid Films. 1978;49(1):3–57. DOI: 10.1016/0040-6090(78)90309-7.

[33] Inoue Y, Nagasawa H, Sone N, Ishino K, Ishida A, Fujiyasu H, Kim J J, Makino H, Yao T, Sakakibara S, Kuwabara M. Fabrication and characterization of short period AlN/GaN quantum cascade laser structures. Journal of Crystal Growth. 2004;265(1-2): 65–70. DOI: 10.1016/j.jcrysgro.2004.01.044.

[34] Ishida A, Ose T, Nagasawa H, Ishino K, Inoue Y, Fujiyasu H. Quantum-cascade structure in AlN/GaN system assisted by piezo-electric effect. Japanese Journal of Applied Physics. 2002;41(2, 3A):236–238. DOI: 10.1143/JJAP.41.L236.

[35] Ishida A, Ose T, Nagasawa H, Inoue Y, Tatsuoka H, Fujiyasu H, Ko H-J, Makino H, Yao T, Kan H. Characterization of AlN/GaN quantum-cascade structures prepared by hot-wall epitaxy. Physica Status Solidi C: Current Topics in Solid State Physics. 2002;0(1): 520–523. DOI: 10.1002/pssc.200390103.

[36] Sigrist M W, editor. Air monitoring by spectroscopic techniques. United States: New York, NY: Wiley-Interscience; 1994. 531 p. DOI: ISBN 0471558753, 9780471558750.

[37] Namjou K, Cai S, Whittaker E A, Faist J, Gmachl C, Capasso F, Sivco D L, Cho A Y. Sensitive absorption spectroscopy with a room-temperature distributed-feedback quantum-cascade laser. Optics Letters. 1998;23(3):219–221. DOI: 10.1364/OL.23.000219.

[38] Faist J, Gmachl C, Capasso F, Sirtori C, Sivco D L, Billargeon J N, Cho A Y. Distributed feedback quantum cascade lasers. Applied Physics Letters. 1997;70(20):2670–2673. DOI: 10.1063/1.119208.

[39] Sharpe S W, Kelly J F, Hartman J S, Gmachl C, Capasso F, Sivco D L, Baillargeon J N, Cho A Y. High-resolution (Doppler-limited) spectroscopy using quantum-cascade

distributed-feedback lasers. Optics Letters. 1998;23(17):1396–1398. DOI: 10.1364/OL.23.001396.

[40] Williams R M, Kelly J F, Hartman J S, Sharpe S W, Taubman M S, Hall J L, Capasso F, Gmachl C, Sivco D L, Baillargeon J N, Cho A Y. Kilohertz linewidth from frequency-stabilized mid-infrared quantum cascade lasers. Optics Letters. 1999;24(24):1844–1866. DOI: 10.1364/OL.24.001844.

[41] Kosterev A A, Curl R F, Tittel F K, Gmachl C, Capasso F, Sivco D, Baillargeon J N, Hutchinson A L, Cho A Y. Effective utilization of quantum-cascade distributed-feedback lasers in absorption spectroscopy. Applied Optics. 2000;39(24):4425–4430. DOI: 10.1364/AO.39.004425.

[42] Kosterev A A, Curl R F, Tittel F K, Gmachl C, Capasso F, Sivco D L, Baillargeon J N, Hutchinson A L, Cho A Y. Absorption spectroscopy with quantum cascade lasers. Laser Physics. 2001;11(1):39–49. DOI: PMID 12143896.

[43] Paldus B A, Harb C C, Spence T G, Zare R N, Gmachl C, Capasso F, Sivco D L, BaillargeonJ N, Hutchinson A L, Cho A Y. Cavity ringdown spectroscopy using mid-infrared quantum-cascade lasers. Optics Letters. 2000;25(9):666–668. DOI: 10.1364/OL.25.000666.

[44] Paldus B A, Spence T G, Zare R N, Oomens J, Harren F J M, Parker D H, Gmachl C, Cappasso F, Sivco D L, Baillargeon J N, Hutchinson A L, Cho A Y. Photoacoustic spectroscopy using quantum-cascade lasers. Optics Letters. 1999;24(3):178–180. DOI: 10.1364/OL.24.000178.

[45] Nägele M, Hofstetter D, Faist J, Sigrist M W. Mobile laser photoacoustic spectrometer for multicomponent trace-gas monitoring based on CO_2- and quantum-cascade lasers as pump sources. In: Conference on Lasers and Electro-Optics Europe; 10–15 September, 2000; Nice, France. IEEE; 2000. p. 12–22. DOI: 10.1109/CLEOE.2000.909779.

[46] Samman A, Rimai L, McBride J R, Carter R O, Weber W H, Gmachl C, Capasso F, Hutchinson A L, Sivco D L, Cho A Y. Potential use of near, mid and far infrared laser diodes in automotive LIDAR applications. In: 52nd Vehicular Technology Conference; 24–28 September, 2000; Boston. USA. IEEE; 2000. p. 2084–2089, vol. 5. DOI: 10.1109/VETECF.2000.883239.

[47] Gittins C M, Wetjen E T, Gmachl C, Capasso F, Hutchinson A L, Sivco D L, Baillargeon J N, Cho A Y. Quantitative gas sensing by backscatter-absorption measurements of a pseudorandom code modulated $\lambda \sim 8$-μm quantum cascade laser. Optics Letters. 2000;25(16):1162–1164. DOI: 10.1364/OL.25.001162.

[48] Sonnenfroh D M, Rawlins W T, Allen M G, Gmachl C, Capasso F, Hutchinson A L, Sivco D L, Baillargeon J N, Cho A Y. Application of balanced detection to absorption measurements of trace gases with room-temperature, quasi-CW quantum-cascade lasers. Applied Optics. 2001;40(6):812–820. DOI: 10.1364/AO.40.000812.

[49] Hvozdara L, Gianordoli S, Strasser G, Schrenk W, Unterrainer K, Gornik E, Murthy C S S S, Kraft M, Pustogow V, Mizaikoff B, Inberg A, Croitoru N. Spectroscopy in the gas phase with GaAs/AlGaAs quantum-cascade lasers. Applied Optics. 2000;39(36):6926–6930. DOI: 10.1364/AO.39.006926.

[50] Kawase K. Terahertz imaging for drug detection and large-scale integrated circuit inspection. Optics and Photonics News. 2004;15(10):34–39. DOI: 10.1364/OPN.15.10.000034.

[51] Lu J Y, Chen L J, Kao T F, Chang H H, Chen H W, Liu A S, Chen Y C, Wu R B, Liu W S, Chyi J I, Sun C K. Terahertz microchip for illicit drug detection. IEEE Photonics Technology Letters. 2006;18(21):2254–2256. DOI: 10.1109/LPT.2006.883285.

[52] Williams B S, Kumar S, Hu Q, Reno J L. High-power terahertz quantum-cascade lasers. Electronics Letters. 2006;42(2):89–91. DOI: 10.1049/el:20063921.

[53] van Neste C W, Senesac L R, Thundat T. Standoff spectroscopy of surface adsorbed chemicals. Analytical Chemistry. 2009;81(5):1952–1956. DOI: 10.1021/ac802364e.

[54] Röpcke J, Davies P B, Glitsch S, Hempel F, Macherius U, Lang N, Rousseau A, Stancu G D, Weltmann K-D, Welzel S, Zimmermann H. Q-MACS, A compact quantum cascade laser absorption spectroscopy system for plasma diagnostics. In: The 2nd International Workshop on Quantum Cascade Lasers; 6–9 September 2006; Brindisi, Italy. Massachusetts Institute of Technology; 2006. p. 388–395.

[55] Jet Propulsion Laboratory. Mars Science Laboratory [Internet]. Available from: http://mars.jpl.nasa.gov/msl/ [Accessed: 19-05-2016].

[56] University of Massachusetts Lowell Sub-Millimetre Wave Laboratory. Sub-Millimetre Wave Laboratory | UMass Lowell [Internet]. Available from: https://www.uml.edu/Research/STL/ [Accessed: 19-05-2016].

[57] Nelson T R. USAF applications for quantum cascade lasers. In: The 2nd International Workshop on Quantum Cascade Lasers; 6–9 September 2006; Brindisi, Italy. Massachusetts Institute of Technology; 2006. p. 455–465.

[58] Danylov A A, Goyette T M, Waldman J, Coulombe M J, Gatesman A J, Giles R H, Qian X, Chandrayan N, Vangala S, Termkoa K, Goodhue W D, Nixon W E. Terahertz inverse synthetic aperture radar (ISAR) imaging with a quantum cascade laser transmitter. Optics Express. 2010;18(15):16264–16272. DOI: 10.1364/OE.18.016264.

[59] Rochat M, Beck M, Faist J, Oesterle U. Measurement of far-infrared waveguide loss using a multisection single-pass technique. Applied Physics Letters. 2001;78(14):1967–1969. DOI: 10.1063/1.1357444.

[60] Vitiello M S, Scalari G, Williams B, Paolo D-N. Quantum cascade lasers: 20 years of challenges. Optics Express. 2015;23(4):5167–5182. DOI: 10.1364/OE.23.005167.

[61] Faist J, Aellen T, Gresch T, Beck M, Giovannini M. Mid-infrared coherent sources and applications. NATO Science for Peace and Security Series ed. The Netherlands: Springer; 2008. 630 p. DOI: ISBN: 978-1-4020-6463-0.

[62] Gmachl C, Ng H M, Cho A Y. Intersubband absorption in GaN/AlGaN multiple quantum wells in the wavelength range of $\lambda \sim 1.75$–4.2 μm. Applied Physics Letters. 2000;77(3):334–336. DOI: 10.1063/1.126968.

[63] Iizuka N, Kaneko K, Suzuki N, Asano T, Noda S, Wada O. Ultrafast intersubband relaxation ($\leq$150 fs) in AlGaN/GaN multiple quantum wells. Applied Physics Letters. 2000;77(5):648–650. DOI: 10.1063/1.127073.

[64] Yao A Y, Hoffman J, Gmachl C. Mid-infrared quantum cascade lasers. Nature Photonics. 2012;6(7):432–439. DOI: 10.1038/nphoton.2012.143.

[65] Belkin M A, Capasso F. New frontiers in quantum cascade lasers: high performance room temperature terahertz sources. Physica Scripta. 2015;90(11):118002. DOI: 10.1088/0031-8949/90/11/118002.

[66] Lu Q, Wu D, Sengupta, Slivken S, Razeghi M. Room temperature continuous wave, monolithic tunable THz sources based on highly efficient mid-infrared quantum cascade lasers. Scientific Reports. 2016;6:23595. DOI: 10.1038/srep23595.

[67] Burghoff D, Kao T-Y, Han N, Chan C W I, Cai X, Yang Y, Hayton D J, Gao J-R, Reno J L, Hu Q. Terahertz laser frequency combs. Nature Photonics. 2014;8:462–467. DOI: 10.1038/nphoton.2014.85.

[68] Harrer A, Szedlak R, Schwarz B, Moser H, Zederbauer T, MacFarland D, Detz H, Andrews A M, Schrenk W, Lendl B. Mid-infrared surface transmitting and detecting quantum cascade device for gas-sensing. Scientific Reports. 2016;6:21795. DOI: 10.1038/srep21795.

[69] Normand E, Howieson I. Quantum-cascade lasers enable gas-sensing technology. Laser Focus World. 2007;43(4):90–92.

Chapter 2

Electronic Band Structure of Quantum Cascade Laser

Arpan Deyasi

Additional information is available at the end of the chapter

http://dx.doi.org/10.5772/66942

Abstract

Multiple quantum well structure is the subject of theoretical and experimental research over the last two decades due to the possibility of making novel electronic and optoelectronic devices. The phenomenon of resonant tunneling makes it a prime candidate for all tunneling-based quantum devices with one-dimensional confinement. THz laser design using multilayered low-dimensional semiconductor structure is one such example, where miniband formation and its energy difference with lowest quantum state play crucial factor in governing device performance. Quantum cascade laser (QCL) is one of such candidate, which speaks in favor of research using multiple-quantum-well (MQW) structure. In the proposed chapter, transmission coefficient of multiple quantum-well structure is numerically computed using propagation matrix technique, and its density of states is calculated in the presence and absence of electric field applied along the direction of quantum confinement. Absorption coefficient is also calculated for its possible application as optical emitter/detector. Based on the electronic and photonic properties investigated, electronic band structure of the quantum cascade laser (formed using the MQW structure) is computed. Formation of miniband is tailored with variation of external bias is shown

Keywords: multiple quantum well, electronic band structure, quantum cascade laser, miniband, band nonparabolicity, electric field

1. Introduction

Research in the area of nanostructures, more precisely on optoelectronic and photonic devices, has progressed in the last few years as novel and unique properties are emerged, which may substantially be able to solve the complex existing problems in the domain of communication and information processing [1–3]. Owing to the synchronized development in material science and nano-fabrication technique [4, 5], in-depth investigations in both experimental and theoretical ways on semiconductor nanostructures lead to the possible realization of novel electronic [6, 7] and photonic [8, 9] devices. Quantization of energy states in low-dimensional semiconductors along different directions leads to the formation of quantum well, wire or dot, which are characteristically different from their bulk counterparts. It would be very difficult today to imagine solid state physics without semiconductor heterostructures, more precisely double heterostructures. The genesis of semiconductor electronics is from the feasibility of conductivity tuning by external impurities by means of doping. This feature along with the concept of non-equilibrium carrier injection helps to breed the semiconductor industry. Heterostructures provide the potential means for solving the far more general problem of controlling fundamental parameters in semiconductor crystals and devices, such as the width of the bandgap, the effective masses and mobilities of charge carriers, the refractive index and the electron energy spectrum. These structures can be used as infrared detectors [10, 11], which has immense applications in medical [12], defense [13] and communication [14] fields. Characteristics of these IR devices may be studied from absorption coefficient profile, and thus its accurate determination becomes very important from device design point of view.

To understand the electronic and optical properties of the nanostructures, computation of their eigenstates is very essential along with density of states profile [15]. Among the quantum structures, quantum well is extensively researched in the last decade because of computational advantages [16–18] and also eases of fabrication [19]. Multiple quantum well structures are subject to theoretical and experimental research in last two decades because of the resonant tunneling phenomenon [20], and novel photonic devices are already designed as transmitter [21] and receiver [22] applicable in nanoelectronic domain. By virtue of quantum engineering, electronic energy states are tuned by suitable application of external excitation in this complex multilayered structure as per the requirement by tailoring the electronic and optoelectronic properties of quantum heterostructures. One such example is the THz laser design with low-dimensional semiconductor structures [23] with multiple layers, where miniband formation and its energy difference with lowest quantum energy state (lowermost energy band) play crucial factor in governing the device performance [21] as optical transmitter. This is one typical semiconductor laser with room temperature operation at IR range, with good peak output power and CW mode of operation [24]. Operation of this unipolar device is based on quantum tunneling and intraband transitions, and formation of miniband critically depends on layer dimensions, their compositions and periodicity [25]. The layer thickness is essentially responsible for determining the wavelength of emitted radiation, as compared to the other semiconductor-based lasers where bandgap of the material determines the wavelength. Thus, accurate determination of electronic band structure of quantum cascade laser (QCL) is very important for theoretical researchers from the point of view of optoelectronic application.

2. Literature review

Semiconductors form the basis for the most modern information processing devices. Electronic devices, such as diodes, bipolar junction transistors and field effect transistors, drive modern electronic technology. Optoelectronic devices, such as laser diodes, modulators and detectors, drive the optical networks. In addition to devices, semiconductor structures have provided the stages for exploring questions of fundamental physics. Quantum Hall effect and other phenomena associated with many-body effects and low dimensions have been studied in semiconductor structures.

Theoretical and experimental researches are so far carried out on the abrupt semiconductor heterostructure composition, precisely the combination of GaAs and AlGaAs. The popularity of this material system is due to the fact of near-perfect match between the GaAs and AlAs binary components. Interestingly, bandgap of GaAs is not suitable for the application of IR detectors, optical transmitter design at infrared region, long-wavelength optical sources and signal processing devices. Several moderate and narrow gap materials are already emerged out for signal processing applications and also for sensing elements in electrical circuits. To name a few, InSb, InAs and GaSb are the materials on which researchers are conceded, and these are categorized as emerging narrow gap materials. Their acceptance of making quantum devices is primarily due to the higher mobilities, bandgaps correspond to the infrared region of electromagnetic spectrum, saturation drift velocities and very low effective mass. Combination of these properties provides higher quantum confinement energies. Higher eigenenergies lead to the possibility of exhibiting mesoscopic property, that is, Aharonov-Bohm effect and Coulomb Blockade effects at higher temperatures.

One of the most interesting research areas in the low-dimensional semiconductor heterostructure is to study novel electronic and photonic properties, which already initiate some technological revolutions [26]. Quantum well, wire and dot are the examples of nanostructure, where quantized dimension has the order of de-Broglie wavelength. Quantum well is the most popular one, can be coupled with external world by means of resonant tunneling. This phenomenon leads the various researches in the last two decade for low-bias applications. Freire [27] calculated multiple-quantum-well (MQW) laser parameters for a dispersion supported transmission system. At very high transmission speed (20 Gbit/s), optical system is simulated for laser parameters using multiple-quantum-well (MQW) system. For strained layer MQW laser, simulation parameters are precisely estimated through signal intensity modulation (IM) response model in order to obtain response curve for communication application. Recently, Taghavi [28] showed that transistor performance can be improved if isolated well is replaced multiple well system. Significant enhancement in device performances is anticipated when the MQW structure is properly designed. Stöhr et al. [29] experimentally characterized electro-optical modulators and switches based on MQW-structure. It can also be used for making quantum cascade laser (QCL) [30], resonant tunneling diode (RTD) [31], resonant tunneling transistor (RTT) [32], quantum well infrared photodetectors (QWIP) [33], etc., which have greater efficiency compared to their bulk counterparts. QCLs are preferred for broadband tuning, RTD and RTT for digital circuits and negative resistance devices. QWIP is used for night vision and medical diagnostics.

Transmission coefficient needs to be computed for analysis of quantum well structures, and also for studying tunneling probability by various numerical approaches like variational method [34], Airy's function approach [35, 36], finite element method [37], transfer matrix technique (TMT) [7, 38–40] and weighted potential method [41]. A comparative analysis among these methods reveals the fact that TMT is one of the best effective procedures. For multibarrier structure analysis at biasing condition, a theoretical estimation about transmission coefficient [17, 38, 42–45] gives corresponding estimation of eigenstates. Significant theoretical contributions on quantum well multilayered heterostructures deal with resonant tunneling mechanism, which is obtained based on accurate solution of the Schrodinger equation subject to the application of uniform magnetic field or constant electric field applied along the perpendicular direction of confinement or along the direction of confinement. Adopting transfer matrix technique, the transmittivity of the multilayered structure is calculated with incident electron energy in the presence or absence external field. Simulated results show good agreement with other existing model, and also with bound-state energies. Based on these calculations, a new class of resonant tunneling superlattice devices has been designed.

Miller [46] investigated the effect of band nonparabolicity on eigenenergies using Kane's two-band model, later considered as a function of material parameters by Hiroshima [47]. Nonparabolicity is accounted for calculation of transmission coefficient by Palomino-Ovando [48]. He used transfer matrix technique (TMT) with nonparabolic dispersion relations for conduction band electrons, which is incorporated in the modified Schrödinger equation; and 2 × 2 matrices are generated. This method is implemented to the study of quasi-periodic semiconductor heterostructures maintaining Fibonacci series property. Miniband formation is observed and amplitudes of resonant tunneling peaks are measured. Results are compared with that obtained for parabolic case to justify the importance of nonparabolic deviations in $Al_xGa_{1-x}As/GaAs$ superlattices.

Multiple quantum well structure is a low-dimensional semiconductor nanostructure which is coupled through the external world by means of resonant tunneling; and based on this physical insight, several novel electronic [31, 32] and photonic [30–33] devices are fabricated which is far superior to their bulk counterparts. The analysis of resonant tunneling diodes was carried out with submicron-sized mesa structures along with contacts at both the terminals in order to measure the current-voltage characteristics in the bistable regime at room temperature. Investigation reveals the noise-triggered firing of spike-like signals from the RTD's biased dynamically, and they act as universal logic gates with reconfigurable property for mini voltage changes ranging from a few mV at the input coordinates. It has also been shown that the RTD junction can be easily integrated to arrays of multiple inputs and has thus the potential to mimic neurons in nanoelectronic circuits.

RTD is already used with slot antenna. With an offset-fed slot antenna, the device is placed at a specified distance w.r.t. slot center, and used in oscillator with considerable output power (100–200 μW) at frequencies of 430–460 GHz. If the distance from the slot center is kept 45 μm for 100 μm long antenna, then 200 μW output power can be obtained (which was the maximum power) at 443 GHz for a single RTD device with a peak current density of 18 mA/μm^2. This opened the door of possible application of quantum well–based devices for communication application at both transmitting and receiving ends. Simultaneous optimization of antenna length and position of quantum device enhances the output performance. Based on this novel investigation and superior performance, a large-format (640 × 512) voltage-tunable quantum-well (QW) infrared photodetector focal plane array (FPA) is designed for both the mid- and long-wavelength infrared (MWIR and LWIR) bands for the sole purpose of imaging applications. Using MBE technique, 8 well MWIR stack with AlGaAs-InGaAs composition and 16 well LWIR stack with AlGaAs-GaAs system are developed and series connection is made between them which provided excellent voltage-tunable spectral response property. Results show a very well redshift of peak responsivity (4.8–8.4 μm, within IR range) with increase of bias within the applicable limit as determined by commercial IC's.

Biasing effect on the lowest three eigenstate of four-barrier three-well MQW structure assuming realistic band structure has been studied by accounting conduction band nonparabolicity. Only first-order nonparabolic effect is considered, and transfer matrix technique is used to solve the problem. GaN/Al_xGa_{1-x} N material composition is assumed for simulation purpose, as the device is aimed to design optical transmitter. The method is well established [38, 39], and results obtained by using TMT are quite accurate when compared with the results obtained using other methods.

Semiconductor nanostructure is the topic of research in the last decade due to its possible novel applications in the field of electronics [9] and photonics [49]. It has been studied that the photonic crystals composed of at least two different semiconductor heterostructures exhibit resonant photonic states, and the composition may be termed photonic double quantum well (PDQW). The heterostructure is schematically denoted as ....Q/P/Q/P/Q...., where P and Q act as photonic wells and barriers depending on their refractive indices, and thus bandgaps, respectively. The resulting band structure exhibits photonic bandgap, similar to the electronic bandgap in semiconductor and dielectric, and photons having energy within that electromagnetic forbidden region could not propagate. This leads to the discrete quantized states. Mathematical modeling has already been carried out for the transmission coefficient of the PDQW heterostructure using both transfer matrix method and plane wave propagation method, and it has been found that within the photonic wells, resonant states exist in split pair mode. This is due to the fact that coupling between degenerate states due to the propagation of forward and backward waves inside this electromagnetic structure, resonant states originate, and it is shared by the both layers of the unit block. Simulation study reveals that resonant states appear as single peak in the transmission spectrum when the energy is within the range of bound photonic state, and then separation distance between the adjacent photonic quantum wells is very large. However, with reduction of the thickness of the sandwiched barrier layer, energy-splitting effect is observed, which is represented by two adjacent resonant peaks in the transmission profile. This realization makes the possible operation of the device as switch single and double transparent states. Similar to the electronic MQW system, in the photonic structure also, number of resonant states can be tailored by varying the dimension of the photonic wells and barriers. Also by increasing the contact barrier widths, quality factor of transmitted peaks can be significantly improved. Researchers are already performed to utilize this property in developing optical filter, transmitter, switch, quantum information processing, receiver and devices, which are suitable in all-optical integrated circuits. The major unique feature if these devices are they are pure photonic devices compared with the existing optoelectronic devices. Quantum confinement is made by reducing the device dimensions of the order of electron wavelength along one, two or three dimensions; results in quantum well, wire or dot, respectively. These quantum systems are usually coupled with the external world through tunneling barriers, which vividly reflect the dominance of the quantum effects to understand the physical properties of heterostructure devices; and the discrete electronic states become resonant one.

Calculation of these eigenstates thus becomes very essential to study the electronic and opto-electronic behavior of the nanodevices [50, 51]. For determining eigenstates, BenDaniel-Duke boundary condition is incorporated in order to consider the effect of effective mass mismatch at different hetero-interfaces, as well as to consider the conduction band discontinuity, which leads to the quantum well potential height. Difference of bandgap gives the second consideration, whereas numerical calculation reveals that results are erroneous when constant effective mass is considered. This signifies the effect of boundary condition at numerical estimation. Structural parameters (in terms of thicknesses of the layers and number of layers) are important players in tuning the energy values. Simulations are also performed in the presence and absence of electric field. In most of the literatures, $GaAs/Al_xGa_{1-x}As$ material composition is considered to understand tunneling probability.

Material compositions of the heterostructures also have a profound effect on the transmission characteristics under resonance condition. Among the quantum structures, quantum well is already studied a lot [16–52] because of computational advantages for analysis and also eases of fabrication [17, 53]. Density of states is one of the very important parameter of nanostructure as it provides the eigenstates from lowermost value [54]. Its profile is changed by application of external field; thus a comparative study my reveal the possibility of low-bias applications through resonant tunneling. Work on density of states of quantum well is carried out by researchers in presence of inelastic scattering [55] along with computation of transmission probabilities. It has been shown that shot noise in a resonant tunneling diode biased in the negative differential resistance regions of the I–V characteristic is enhanced with respect to 'full' shot noise. Experimental results suggests that profile of density of states function for particular type of quantum well system is the governing cause of the so-called noise effect, physically controlled by electron-electron interaction through Coulomb force. Numerical results are also presented from the above-mentioned theory, which are in close proximity with the experiment. This ensures that the model presented is accountable for the relevant physics involved in the phenomenon.

Step-like shape of the density of state (DOS) profile is used to study the characteristics of heterostructure well laser [56]. For finite-barrier quantum well, it differs from what obtained for infinite well problem [57]. The results presented in this chapter, obtained by applying the correspondence with the bulk density of states (DOS), indicate that the DOS in finite-barrier quantum wells differs substantially from the conventional 2D DOS calculated for infinite-barrier quantum wells.

Peak of the DOS is analytically related with the occurrence of resonance [58] along with transmission and reflection phase times. Existence of quasi-bound states is observed from DOS plot [59] for specially designed rectangular well structures. Numerical calculations have been presented, which substantially prove that electron states are present in the continuous part of the spectrum for different quantum well devices irrespective of well geometry. Initial works are carried out on rectangular potential profiles. Necessary boundary conditions for creation of these quasi-states are theoretically investigated. At very high energies in conduction band, existence of these quasi-bound states with spatial localization is determined. The results can be very useful in design of semiconductor optoelectronic devices whose working principle will be based on continuum bound states.

With the emergence of nanoelectronic devices as the possible replacement of existing VLSI technology, the need of resonant tunneling device in sub-micron domain becomes very important for low bias application point of view, and theoretical research for accurate design of these structures leads to the invention of novel electronic and optoelectronic properties. Ultrathin semiconductor quantum devices are now the interests of research where quantum well, wire and dot have already realizable because of the advancement of microelectronics technology. For accurate estimation of resonant tunneling, displacement of energy levels from the band-edge should be considered; hence, realistic band structure plays a very important role for different material parameters. This leads to the better computation of intersubband transition energy for optical emitter/detector design [50, 60]. Intersubband transition energy is computed for core-shell (normal and inverted) quantum dots (CSQD) of cubic and spherical geometries by solving time-independent Schrödinger equation using finite-difference technique. Sparse, structured Hamiltonian matrices of order $N^3 \times N^3$ for cubic and $N \times N$ for spherical dots are produced considering N discrete points in spatial direction. The matrices are diagonalized to obtain eigenstates for electrons. Computed results for the lowest three eigenstates and intersubband transitions are shown for different structural parameters taking $GaAs/Al_xGa_{1-x}As$-based CSQD as an example. Reduction of subband transition energy is observed with increase in core thickness. A comparative study reveals that spherical CSQD exhibits higher transition energy between any two subbands than that is demonstrated by cubic CSQD of similar dimension and equal material composition. More interesting phenomenon is observed for inverted core-shell structure. In this case, transition energy monotonically decreases for cubic dot, while it is increased for spherical dot with enhancement in core size. These devices are utilized as optical emitter/detector due to the wider range of tuning in intersubband transition through tailoring dot dimensions.

Miller investigated the effect of conduction band nonparabolicity on eigenstates using Kane's two band model. Hiroshima showed the importance of material parameters to evaluate the same. Nelson introduced the energy-dependent effective mass theory [61] following Ben-Daniel Duke boundary condition for precise estimation of carrier energies. Palomino-Ovando extended the analysis for superlattice structure [62] where energy was calculated form transmission coefficient peaks. The work of Li in determining the electron states in multilayer semiconductor quantum devices plays a significant role. He showed that finite element method can effectively be used for near accurate determination of these electron states. Moreover, he emphasized on the importance of band structure consideration and showed the effect of Kane type conduction band nonparabolicity over the carrier states. For two different material compositions, the result is presented in tabular form.

This work is incorporated in modulator design [63]. Using multiple quantum well structures, authors show that spatial light modulators can be designed. Reflectance levels are calculated for 128 × 128 pixels ternary spatial light modulators with 10 KHz operating frequency. It is shown that single reflectance level can be duo with just a 180° phase difference. Structural design parameters of the modulator will play important role in determining the performance of the device.

Due to the rapid advancement in microelectronics technology [64, 65] in the last decade, research in the domain of semiconductor nanostructure has been carried out to a great extent by both theoretical [4, 66] and experimental workers [5, 6]. Quantum well waveguide is fabricated by ion implantation technique using InGaAsP/InP material composition. Multilayer structure is developed for waveguide gratings precisely for the application at CWDM wavelengths. Reactive ion etching technique is utilized to reduce the roughness below 5 nm (experimental results show surface roughness is approximately 2–3 nm). Significant findings emphasize the importance of this quantum device as cross talk between adjacent channels goes below −10 dB level in the transmission spectra of the waveguide. Moreover, insertion loss of the grating becomes less than 5 dB, which is another major advantage for communication application.

Heterostructure light emitting diode is made by InGaN/GaN composition using MOCVD technology. Strucutre is developed over $LiAlO_2$ (100) substrates consisting of multiple quantum well with five wells along with GaN p-n junction. Colors of the output emission (generally blue or green) depend on the indium composition of InGaN layer of the MQW structure. Proof of quantum well interface is experimentally obtained by third-order satellite peaks. These are measured by high resolution x-ray diffraction set-up. Good performances are reported in various literatures; for example, 3 mW output power is achieved for $800 \times 800\ \mu m^2$ blue LED device with 200 mA input current. It is also experimentally observed that for green and blue LED's with InGaN/GaN composition, electroluminescence (EL) spectra suggest that peak wavelength almost saturates at some precise positions of the spectrum with enhancement of injection current. Henceforth, it may very easily be concluded that polarization fields are totally absent in the active region of the structure.

Spin-photonic semiconductor devices are the another novel class of quantum well based devices, which are researched in recent days due to the ability of controlling electronic and optoelectronic properties using the spin of conduction electrons. The two most promising examples of these devices are spin-switches and spin-vertical-cavity surface-emitting lasers (spin-VCSELs). Experimental investigation showed that circularly polarized laser with very high degree of circular polarization (0.96) at room temperature is obtained using GaAs/AlGaAs multiple quantum well vertical cavity surface emitting laser structure where spin relaxation time is very high (0.7 ns). Furhter works with the same structure also showed that carrier lifetime and electron spin relaxation time can be tuned by design parameters. This also modifies the lasing characteristic. Orientation of the wafer plane plays a very crucial role in this regard. All the experimental results are obtained for <110> wafer. Investigation is also conducted on p-i-n structure with similar material composition. Result showed that remarkable reduction of spin relaxation time can be obtained (4–0.3 ns) when the quantum well structure is subjected under external electric field applied along the direction of well width at room temperature. To understand the electronic and optical properties of the nanostructures, computation of their eigenstates is very essential along with density of states profile [16]. Among the quantum structures, quantum well is extensively researched in the last decade because of computational advantages [7, 8, 67] and also eases of fabrication [9].

Transfer matrix technique (TMT) and propagation matrix method (PMM) are the two most suitable cost-effective methodologies for the analysis of wave function profile, confined energy states and transmission coefficients in superlattice nanostructures. There are other numerical methods like finite element method (FEM), finite difference method (FDM) but the complexity of the programming makes the computational time larger. However, these later methods are very accurate regarding the precision of the obtained results. Material composition for the ternary or quaternary heterostructures plays the bigger role along with Ben-Daniel Duke boundary conditions for determining the basic electrical properties of multilayered structures. Structural parameters within fabrication limit showed the variation of electronic (eigenenrgy, density of states, for example) and optoelectronic (absorption coefficient, oscillator strength, for example) properties of this superlattice structure which are pivotal in analysis of further complex structures. Otherwise multiple quantum well structures become a superlattice structure when width of the barrier layer between adjacent quantum wells is reduced to very small dimension, and this ensures the beginning of resonant tunneling through intermixing of the wavefunctions between the quantum wells. This provides a new insight in the nature of interacting wave functions for thin barriers. Number of wells in this computation becomes also important along with the structural parameters and material compositions. Different iterative methods are implemented for complex geometrical structures. Doping of the wells is taken into account through self-consistent solution of Schrödinger and Poisson's equations.

Thus, a comprehensive analysis of electronic and optical properties of MQW structures will make the foundation for understanding the quantum cascade laser. Its principle of operation is essentially dependent on the band structure, layer thickness is primarily responsible for determining the wavelength of emitted radiation, as compared to the other semiconductor based lasers where bandgap of the material determines the wavelength. Thus accurate determination of electronic band structure of quantum cascade laser is very important for theoretical researchers from the point of view of optoelectronic application.

3. Transmission coefficient of multiple quantum well structure

For analyzing electrical property of MQW structure, first transmission coefficient is calculated by considering a three-well four-barrier structure with rectangular potential profile configuration having GaN as well layer and $Al_xGa_{1-x}N$ as barrier layer. It is observed that by increasing well width, transmission probability increases. This is due to the fact that by increasing tunneling dimension of the well, quantum confinement decreases, so transmission can be achieved at lower energy values. By changing the material composition of barrier material, it is observed that transmission coefficient decreases with increase of Al mole fraction. It is plotted in **Figure 1**. This is due to the fact that by increasing Al percentage, potential height increases and effective mass mismatch at the junction also increases. This increases quantum confinement, which increases the eigenenergy. By increasing the thickness of the contact barrier, transmission probability reduces, whereas it increases if contact barrier thickness is reduced compared to that of the internal barriers. This is shown in **Figure 2** for both parabolic and nonparabolic band structures. By increasing number of wells, it is observed the eigenvalue reduces, and ultimately becomes constant for higher numbers, as evident from **Figure 3**. Eigenvalue increases with increase of material composition, as suggested from previous results, plotted in **Figure 4**. Obviously, for nonparabolic band structure, eigenenergy is less compared to the ideal parabolic concept. The variation is almost linear, and the gap reduces with increase of al mole fraction.

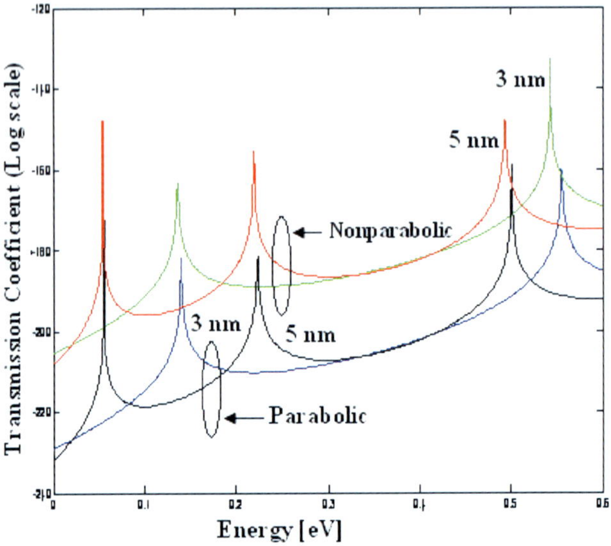

Figure 1. Transmission coefficient with energy for different barrier material composition for both parabolic and nonparabolic structure.

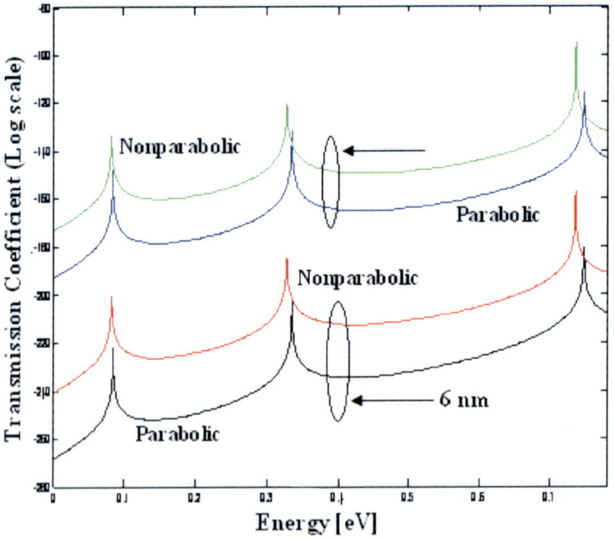

Figure 2. Transmission coefficient with energy for asymmetric barrier width for both parabolic and nonparabolic structure.

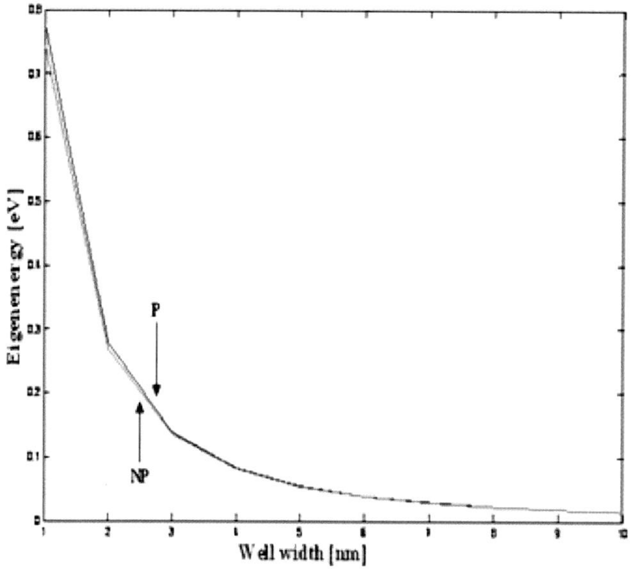

Figure 3. Eigen energy with dimension of well for both parabolic and nonparabolic structure.

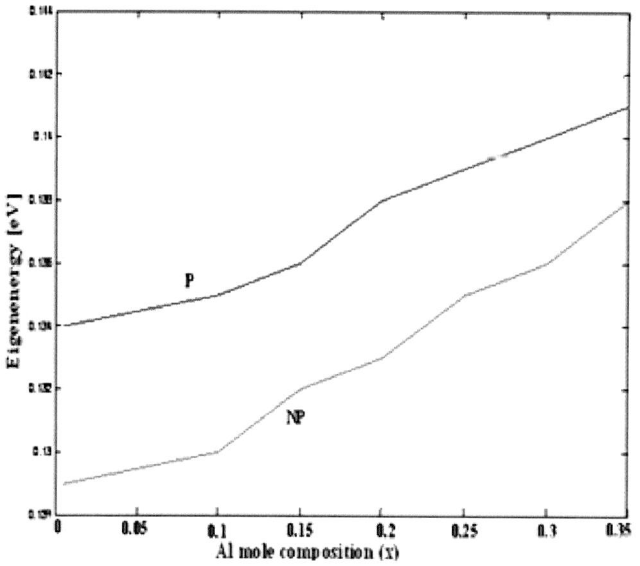

Figure 4. Eigen energy with barrier material composition for both parabolic and nonparabolic structure.

4. Density of states of MQW structure

In **Figure 5**, density of states (DOS) is plotted for lowest two eigenstates with different well widths. It is observed from the plot that by increasing the well width, eigenenergy appears at lower energy values due to the reduction of quantum confinement. It may also be seen that higher bandgap system provides eigenstate at lower energy range. Hence, optical device based on GaN/AlGaN system can be tuned at lower bias. Also due to closeness of first two energy levels in this composition, intersubband transition energy is higher for InAs/GaInAs system, which makes it efficient candidate for high frequency laser.

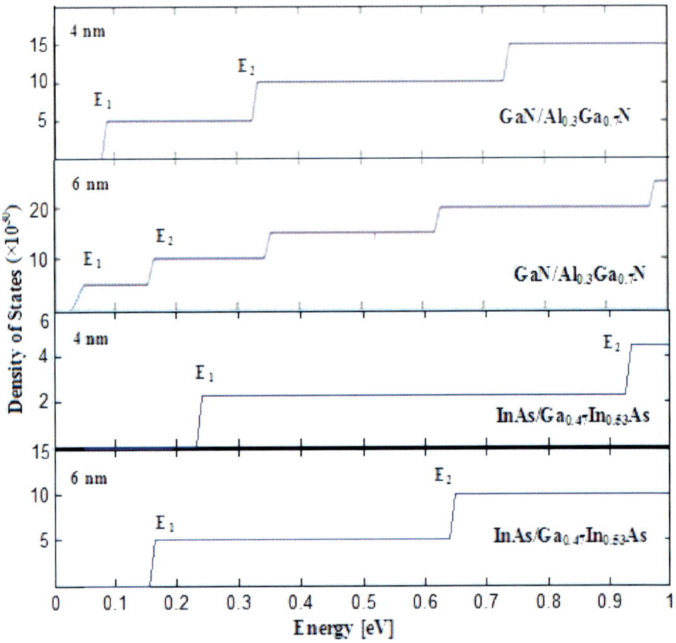

Figure 5. Density of states for the lowest two eigenstates for different well widths and different material compositions with nonparabolic dispersion relation.

Figure 6 shows the comparative study of DOS in presence and absence of electric field for InAs/GaInAs material composition. From the graph, it may be noted that application of transverse field lowers the magnitude of eigenstates irrespective of dispersion relation considered for simulation. Results are also compared with overestimated parabolic assumption. It may be observed from the comparison that incorporation of band nonparabolicity reduces the eigenvalue. Hence, for fine wavelength tuning purpose to design photonic transmitter/detector, realistic band structure consideration plays a vital role.

http://dx.doi.org/10.5772/66942

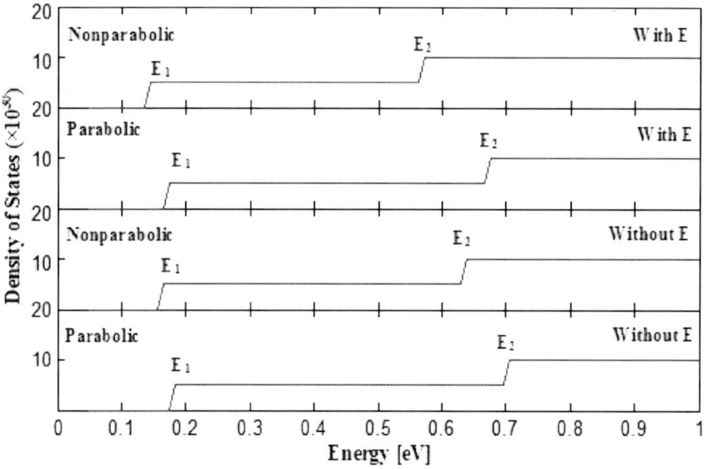

Figure 6. Density of States for the lowest two eigenstates in presence of absence of electric field for nonparabolic and parabolic structures.

By varying the material composition of barrier layers, it is observed that quantum states appear in higher energy values with increase of GaN mole fraction. The result is shown in **Figure 7**. This is because with increase of x, mismatch of effective mass increases as well as the conduction band discontinuity. This enhances the quantum confinement. Hence, eigenenergy increases. This is reflected via density of states plot.

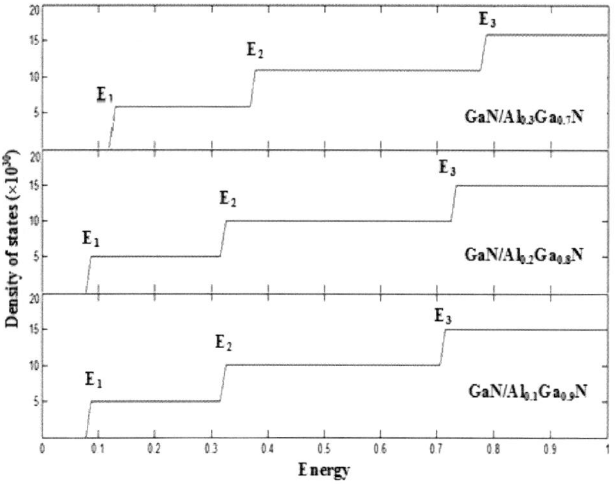

Figure 7. DOS of superlattice with different material compositions in absence of field considering band nonparabolicity.

5. Absorption coefficient of MQW structure

Absorption coefficient is calculated as a function of operating wavelength for different well width, as shown in **Figure 8**. For lower well dimension, it is found that the absorption coefficient value is less compared to the higher well dimension. Again, when the energy difference between the bands increases, interband transition energy increases. This reduces the peak of absorbance amplitude, provided half-width at half-maximum is kept constant throughout the simulation.

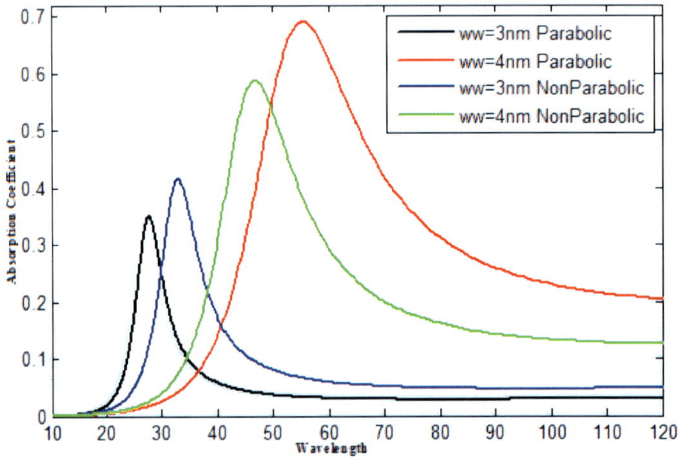

Figure 8. Absorption coefficient for different well dimension with wavelength variation for parabolic, nonparabolic dispersion.

With increase of well dimension, redshift is observed for all the profiles corresponding to the transition energies; also peak value is shifted rightwards and the wavelength value increases. When the parabolic and nonparabolic dispersion relation is considered for same well dimension the peak value is greater for lower well dimension nonparabolic case. But with change in eigenenergy, for higher well dimension the parabolic relation gives a higher absorption peak value. Thus the nonparabolic dispersion relation takes the priority for simulation.

With parabolic and nonparabolic relations into consideration the material composition is varied in accordance with the wavelength. The absorption coefficient profiles are depicted in **Figure 9**. For parabolic dispersion, the transition energies are low, so the absorption peak values are less compared to that of nonparabolic case. Also for higher material composition the shift in absorption value is considerable. The peak absorption values are greater for higher material composition and also for nonparabolic dispersion.

http://dx.doi.org/10.5772/66942

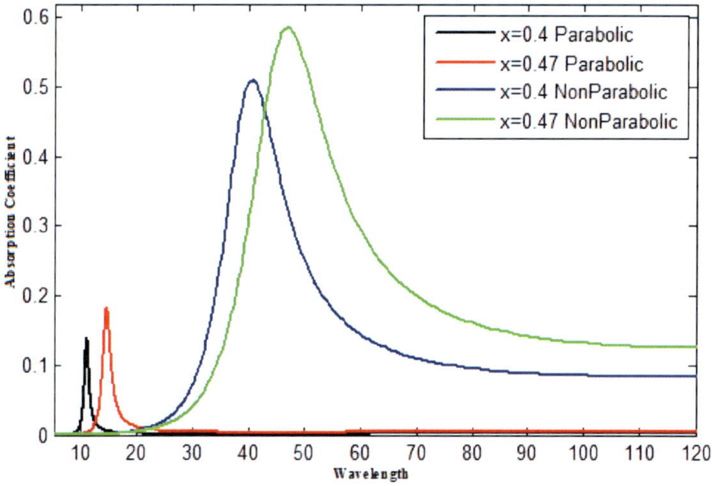

Figure 9. Absorption coefficient for different material composition with wavelength variation for nonparabolic and parabolic dispersion.

6. Band structure of quantum cascade laser

Electronic band structure of quantum cascade laser is analytically computed in presence of electric field applied along the direction of quantum confinement. At first, electronic and optoelectronic properties of the MQW structure is calculated. Next, miniband formation is observed for quantum cascade laser for a few precise magnitudes of electric fields. Separation of the miniband w.r.t. lowest energy band is calculated [68]. $Al_xGa_{1-x}As/GaAs$ composition is taken into account with the incorporation of Ben-Daniel Duke boundary conditions at hetero-interfaces. Eigenstates are calculated in presence and absence of Kane-type first-order conduction band nonparabolicity. Its variation as a function of applied field is determined, and is compared with the findings of zero bias conditions.

Figures 10–12 represent the electronic band structure of quantum cascade laser. For biasing, electric field is applied along the direction of quantum confinement, and eigenstates are computed. Results are also obtained when field is absent. While designing the structure, the most important aspect is that a distinguishable separation between injector and active region should be kept along with the miniband formation [68]. **Figure 10** shows the band diagram when external field is totally absent, and **Figure 11** demonstrated it for very high electric field (56.8×10^5 V/m). Here it may be noted that all the high magnitudes of electric field will not provide desirable QCL operation, as discrete miniband formation is necessary for that purpose. From the simulated observations, it may be indicated that the eigenenergy states are totally discrete in absence of external excitation, which are nothing but the eigenstates of a simple multiple quantum-well (MQW) structure.

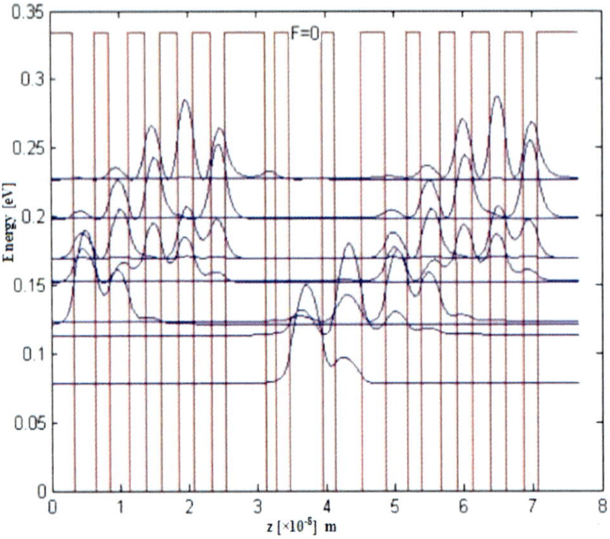

Figure 10. Electronic band structure of QCL without external bias.

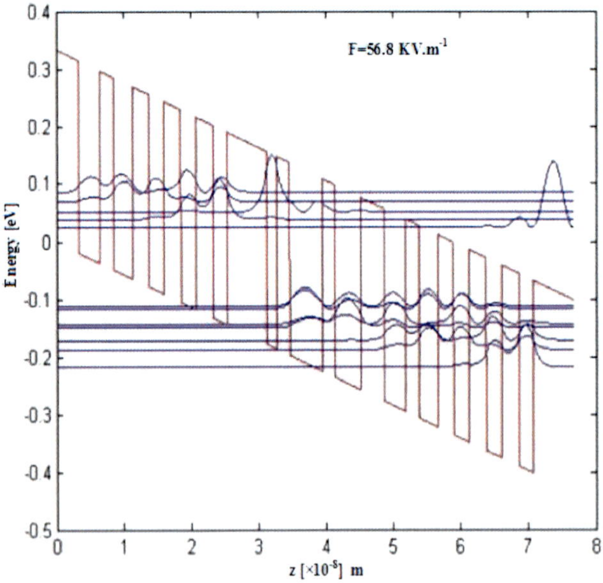

Figure 11. Electronic band structure of QCL with 56.8×10^5 V/m.

From **Figure 11**, it is observed that after application of very high electric field, miniband is formed over the ground state energy band. In this context, it may be noted down that after the injector region, active region starts, and the wavefunction starts to produce minibands. The energy gap between miniband and ground state energy band is distinguishable, which ensures the QCL operation. In this context, it may be pointed out that the magnitude of external field is chosen based on the structural dimensions and the material composition taken for the simulation. The structure is titled compared to **Figure 10**, along the direction of electric bias. Thus, the periodic growth of wavefunction in the miniband appears outside the confinement region, that is, in the quasi-continuous region. This band structure modulation is absent if the field is moderate (35.8 × 10^5 V/m).

Further reduction of electric field makes position of the miniband almost inside the confinement region, that is, miniband position is below the quasi-continuous region. This is shown in **Figure 12**. Eigenenergy variation with the applied bias for the laser is shown in **Figure 13**. Simulation result suggests that with increase of applied field, eigenenergy increases monotonically, and the rate reduces once filed crosses the value 35 × 10^5 V/m. Thus, stimulated emission between the miniband and the ground state energy band is effectively controlled by external bias.

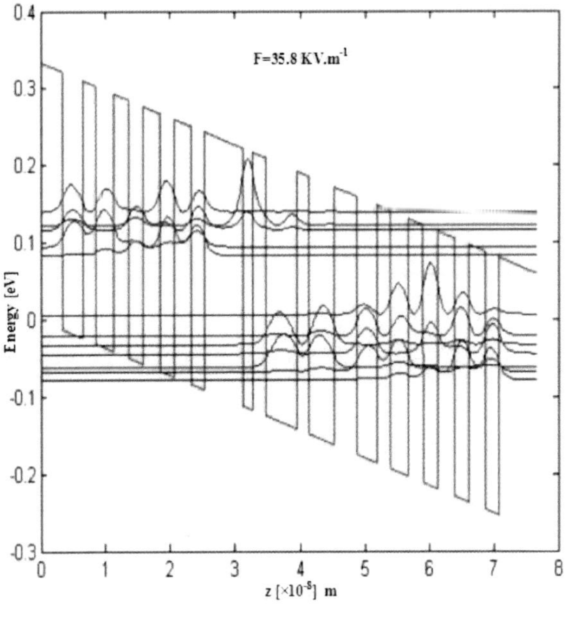

Figure 12. Electronic band structure of QCL with 35.8 × 10^5 V/m.

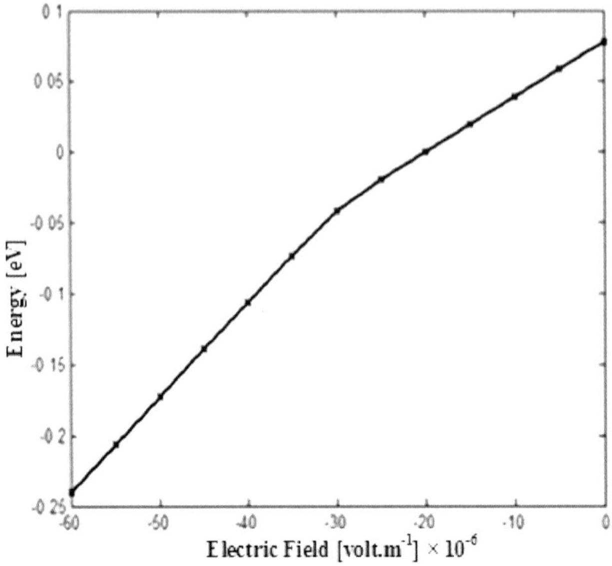

Figure 13. Variation of eigenenergy with applied field.

7. Conclusion

In this article, a detailed investigation on electrical and optoelectronic properties of multiple quantum well structure is carried out. Based on that structure, wavefunction in a quantum cascade laser at various biased conditions is analytically calculated and also for unbiased condition. It is proven and shown that miniband formation is only possible at some precise electric field. The method showed can also be applicable for the other structures. Key factor in this calculation is that position-dependent effective mass is considered for the simulation.

Author details

Arpan Deyasi

Address all correspondence to: deyasi_arpan@yahoo.co.in

Department of Electronics and Communication Engineering, RCC Institute of Information Technology, Kolkata, India

References

[1] Basu P K, Ghosh S, Mukhopadhyay B, Sen G: Ge/SiGeSn multiple quantum well photonic devices. 2nd National Workshop on Advanced Optoelectronic Materials and Devices. US, 2008.

[2] Fara S, Sterian P, Fara L, Iancu M, Sterian A: New results in optical modeling of quantum well solar cells. International Journal of Photoenergy. 2012, 2012:810801.

[3] Friel I, Thomidis C, Moustakasa T D: Ultraviolet electroabsorption modulator based on AlGaN/GaN multiple quantum wells. Journal of Applied Physics. 2005, 97:123515.

[4] Zhong S, Qu X: Design and fabricate InGaAlAs quantum well device for future optoelectronic integration. Advanced Materials Research. 2012, 442:188–192.

[5] Liu B, Han P, Xie Z, Zhang R, Liu C, Xiu X, Hua X, Lu H, Chen P, Zheng Y, Zhou S: Fabrication of blue and green non-polar InGaN/GaN multiple quantum well light-emitting diodes on LiAlO2(100) substrates. Physica Status Solidi(a). 2010, 207:1404–1406.

[6] Kuo Y, Lee Y K, Ge Y, Ren S, Roth J E, Kamins T E, Miller D A B, Harris J S: Strong quantum-confined Stark effect in germanium quantum-well structures on silicon. Nature. 2005, 437:1334–1336.

[7] Samuel E P, Patil D S: Analysis of wavefunction distribution in quantum well biased laser diode using transfer matrix method. Progress in Electromagnetics Research Letters. 2008, 1:119–128.

[8] Hitoshi K: Spin-photonic semiconductor devices based on (110) quantum wells: Spin-VCSELs and spin-switches. 13th International Conference on Transparent Optical Networks. IEEE. 2011, 1–4.

[9] Joel D, Singh M R: Resonant tunneling in photonic double quantum well heterostructures. Nanoscale Research Letters. 2010, 5:484–488.

[10] Yu C H, Zhang B, Lu W, Shen S C, Liu H C, Fang Y Y, Dai J N, Chen C Q: Strong enhancement of terahertz response in GaAs/AlGaAs quantum well photodetector by magnetic field. Applied Physics Letters. 2010, 97:022102.

[11] Kalchmair S, Detz H, Cole G D, Andrews A M, Klang P, Nobile M, Gansch R, Ostermaier C, Schrenk W, Strasser G: Photonic crystal slab quantum well infrared photodetector. Applied Physics Letters. 2011, 98:011105.

[12] Perera A G U: Quantum structures for multiband photon detection. Opto-Electronics Review. 2006, 14:103–112.

[13] Phillips M C, Taubman M S, Bernacki B E: Design and performance of a sensor system for detection of multiple chemicals using an external cavity quantum cascade laser. Proceedings of SPIE. 2010, 7608:76080D.

[14] Wei R, Deng N, Wang M, Zhang S, Chen P, Liu L, Zhang J: Study of Self-assembled Ge Quantum Dot Infrared Photodetectors. 1st IEEE International Conference on Nano/Micro Engineered and Molecular Systems. US, 2006, 330–333.

[15] Harrison P: Quantum Wells, Wires and Dots: Theoretical and Computational Physics of Semiconductor Nanostructures, John Wiley, New York, 4th edition, 2016.

[16] Talele K, Patil D S: Analysis of wave function, energy and transmission coefficients in GaN/AlGaN superlattice nanostructures. Progress in Electromagnetics Research. 2008, 81:237–252.

[17] Simion C E, Ciucu C I: Triple-barrier resonant tunneling: A transfer matrix approach. Romanian Reports in Physics. 2007, 59:805–817.

[18] Kundu P, Ghosh P, Deyasi A: Analytical computation of absorption coefficient for intersubband transition in MQW structure. Lecture Notes in Electrical Engineering: Computational Advancement in Communication Circuits and Systems [SPRINGER], part 6. Advances in Devices and Circuit. 2014, chapter 35, 335:321–329.

[19] Andrew S R, Marsh J H, Holland M C, Kean A H: Quantum well laser with integrated passive waveguide fabricated by neutral impurity disordering. IEEE Photonics Technology Letters. 1992, 4:426–428.

[20] Kubis T, Mehrotra S R, Klimeck G: Design concepts of terahertz quantum cascade lasers: Proposal for terahertz laser efficiency improvements. Applied Physics Letters. 2010, 97:261106.

[21] Cooper J D, Valavanis A, Ikonic Z, Harrison P, Cunningham J E: Finite difference method for solving the Schrödinger equation with band nonparabolicity in mid-infrared quantum cascade lasers. Journal of Applied Physics. 2010, 108:113109.

[22] Lin S H, Feng D J Y, Lee M L, Lay T S, Sun T P, Kuan C: Double-barrier superlattice infrared photodetector integrated with multiple quantum-well infrared photodetector to improve performance. International Journal of Electrochemical Science. 2012, 7:5746–5753.

[23] Razavipour S G, Dupont E, Chan C W I, Xu C, Wasilewski Z R, Laframboise S R, Hu Q, Ban D: A high carrier injection terahertz quantum cascade laser based on indirectly pumped scheme. Applied Physics Letters. 2014, 104:041111.

[24] Bugajski M, Kosiel K, Szerling A, Karbownik P, Pierściński K, Pierścińska D, Hałdaś G, Kolek A: High performance GaAs/AlGaAs quantum cascade lasers: Optimization of electrical and thermal properties. Proceedings of SPIE. 2012, 8432:84320I.

[25] Yamamoto H: Resonant tunneling condition and transmission coefficient in a symmetrical one-dimensional rectangular double-barrier system. Applied Physics A: Materials Science & Processing. 1987, 42:245–248.

[26] ITRS: Roadmap, http://www.itrs2.net/itrs-reports.html, 2007.

[27] Freire M, Silva H J A: Estimation of multiple-quantum well laser parameters for simulation of dispersion supported transmission systems at 20 Gbit/s. IEE Proceedings-Optoelectronics. 1999, 146:93–98.

[28] Taghavi I, Kaatuzian H, Leburton J P: Multiple versus single quantum well transistor laser performances. Integrated Photonics Research, Silicon and Nanophotonics. US, 2012, IM4B: IM4B.5

[29] Stöhr A, Humbach O, Zumkley S, Wingen G, David G, Jäger D, Bollig B, Larkins E C, Ralston J D: InGaAs/GaAs multiple-quantum-well modulators and switches. Optical and Quantum Electronics. 1993, 25:S865–S883.

[30] Fujita K, Edamura T, Furuta S, Yamanishi M: High-performance, homogeneous broad-gain quantum cascade lasers based on dual-upper-state design. Applied Physics Letters. 2010, 96:241107.

[31] Worschech L, Hartmann F, Kim T Y, Ho-fling S, Kamp M, Forchel A, Ahopelto J, Neri I, Dari A, Gammaitoni L: Universal and reconfigurable logic gates in a compact three-terminal resonant tunneling diode. Applied Physics Letters. 2010, 96:042112.

[32] Hinata K, Shiraishi M, Suzuki S, Asada M, Sugiyama H, Yokoyama H: Sub-terahertz resonant tunneling diode oscillators with high output power (200 μw) using offset-fed slot antenna and high current density. Applied Physics Express. 2010, 3:014001.

[33] Eker S U, Kaldirim M, Arslan Y, Besikci C: Large-format voltage-tunable dual-band quantum-well infrared photodetector focal plane array for third-generation thermal imagers. IEEE Electron Device Letters. 2008, 29:1121–1123.

[34] Bastard G, Mendez E E, Chang L L, Esaki L: Variational calculations on a quantum well in an electric field. Physical Review B. 1983, 28:3241–3245.

[35] Ghatak A K, Thyagarajan K, Shenoy M R: A novel numerical technique for solving the one-dimensional Schrödinger equation using matrix approach– application to quantum well structures. IEEE Journal of Quantum Electronics. 1988, 24:1524–1531.

[36] Brennan K F, Summers C J: Theory of resonant tunneling in a variably spaced multi-quantum well structure: An airy function approach. Journal of Applied Physics. 1987, 61:614–623.

[37] Hayata K, Koshiba M, Nakamura K, Shimizu A: Eigenstate calculations of quantum well structures using finite elements. Electronics Letters. 1988, 24:614–616.

[38] Sugg A R, Leburton J P C: Modeling of modulation-doped multiple-quantum-well structures in applied electric fields using the transfer-matrix technique. IEEE Journal of Quantum Electronics. 1991, 27:224–231.

[39] Jonsson B, Eng S T: Solving the Schrödinger equation in arbitrary quantum-well profiles using the transfer-matrix method. IEEE Journal of Quantum Electronics. 1990, 26:11.

[40] Hong Y J, Zhi J G, Yan Z, Wu L W, Chun S Y, Guo W Z, Jun X J: Resonant tunneling in barrier-in-well and well-in-well structures. Chinese Physics Letters. 2008, 25:4391–4394.

[41] Tsuji Y, Koshiba M: Analysis of complex eigenenergies of an electron in two- and three-dimensionally confined systems using the weighted potential method. Microelectronics Journal. 1999, 30:1001–1006.

[42] Austin E J, Jaros M: Electronic structure of an isolated GaAs-GaAlAs quantum well in a strong electric field. Physical Review B. 1985, 31:5569–5572.

[43] Harwit A, Harris J S: Calculated quasi-eigenstates and quasi-eigenenergies of quantum well superlattices in an applied field. Journal of Applied Physics. 1986, 60:3211–3213.

[44] Guoa Y, Gu B L, Yu J Z, Zeng Z, Kawazoe Y: Resonant tunneling in step-barrier structures under an applied electric field. Journal of Applied Physics. 1988, 84:918–924.

[45] Allen S S, Richardson S L: Theoretical investigations of resonant tunneling in asymmetric multibarrier semiconductor heterostructures in an applied constant electric field. Physical Review B. 1994, 50:11693–11700.

[46] Miller R C, Kleinman D A, Gossard A C: Energy-gap discontinuities and effective masses for GaAs-$Al_xGa_{1-x}As$ quantum wells. Physical Review B. 1984, 29:7085–7087.

[47] Hiroshima T, Lang R: Effect of conduction-band nonparabolicity on quantized energy levels of a quantum well. Applied Physics Letters. 1986, 49:456–457.

[48] Palomino-Ovando M, Cocoletzi G H, Perez-Lopez C: Band nonparabolicity in quasi-periodic Fibonacci heterostructures. Physics Letters A. 1996, 213:191–196.

[49] Chanda A, Eastman L F: Quantum mechanical reflection at triangular planar-doped potential barriers for transistors. Journal of Applied Physics. 1982, 53:9165–9169.

[50] Deyasi A, Bhattacharyya S, Das N R: Computation of intersubband transition energy in normal and inverted core-shell quantum dots using finite difference technique. Superlattices & Microstructures. 2013, 60:414–425.

[51] Deyasi A, Bhattacharyya S, Das N R: A finite difference technique for computation of electron states in core-shell quantum wires of different configurations. Physica Scripta. 2014, 89:065804.

[52] Esaki L, Chang L L: New transport phenomenon in a semiconductor superlattice. Physical Review Letters. 1974, 33:495–498.

[53] Tsukamoto S, Nagamune Y, Nishioka M, Arakawa Y: Fabrication of GaAs quantum wires on epitaxially grown V grooves by metal-organic chemical-vapor deposition. Journal of Applied Physics. 1992, 71:533–535.

[54] Arakawa Y, Nagamune Y, Nishioka M, Tsukamoto S: Fabrication and optical properties of GaAs quantum wires and dots by MOCVD selective growth. Semiconductor Science and Technology. 1993, 8:1082–1088.

[55] Khalil H M, Balkan N: Carrier trapping and escape times in p-i-n GaInNAs MQW structures. Nanoscale Research Letters. 2014, 9:21.

http://dx.doi.org/10.5772/66942

[56] Iannaccone G, Pellegrini B: Compact formula for the density of states in a quantum well. Physical Review B. 1996, 53:2020–2025.

[57] Holonyak N, Kolbas R M, Dupuis R D, Dapkus P D: Quantum-well heterostructure lasers. IEEE Journal of Quantum Electronics. 1980, 16:170–186.

[58] Osinski M, Mojahedie M: Density of confined states in finite-barrier quantum wells. LEOS '93 Conference Proceedings. IEEE. 1993, 635–636.

[59] Carrillo-Delgado E A, Rodriguez-Vargas I, Vlaev S J: Continuum electronic bound states in rectangular quantum wells and barriers. PIERS. 2009, 5:137–140.

[60] Deyasi A, Bhattacharyya S: Interband transition energy of circular quantum dots under transverse magnetic field. Physics Procedia. 2014, 54:118–126.

[61] Nelson D F, Miller R C, Kleinman D A: Band nonparabolicity effects in semiconductor quantum wells. Physical Review B. 1987, 35:7770.

[62] Le K Q: Finite element analysis of quantum states in layered quantum semiconductor structures with band nonparabolicity effect. Microwave and Optical Technology Letters. 2009, 51:1.

[63] Junique S, Wang Q, Martijn H H, Guo J, Noharet B, Borglind J, Hirschauer B, Malm H, Agren D, Oeberg O, Andersson J Y: Multiple quantum well spatial light modulators: Design, fabrication, characterization. Proceedings of SPIE 4457, Spatial Light Modulators: Technology and Applications, 62.

[64] Sonkar R K, Das U: Fabrication of F-ion implanted quantum well intermixed waveguide grating. Photonics Global Conference. Singapore, Dec, 2012, 1–3.

[65] Choy C H W, Chan K S: Theoretical analysis of diffused quantum-well lasers and optical amplifiers. Selected Topics in IEEE Journal of Quantum Electronics. 2003, 9:698–707.

[66] Sun Y H, Xu L J, Zhang B, Xu Q F, Wang R, Dai N, Wu H Z: Theoretical analysis of optical gain in $PbSe/Pb_{1-x}Sr_xSe$ quantum well lasers. Physica Status Solidi(a). 2009, 206:2606–2612.

[67] Sarkar D, Deyasi A: Calculating Absorption Coefficient of Gaussian Double Quantum Well Structure with Band Nonparabolicity for Photodetector in Microwave Spectra: Foundations and Frontiers in Computer, Communication and Electrical Engineering. 2016, CRC Press. Taylor & Francis Group, London, UK. chapter 47: 225–229.

[68] Chakraborty R, Deyasi A, Paul A, Nayak S: Electronic band structure of quantum cascade laser. Accepted in Springer Proceedings in Physics: Advances in Optical Science and Engineering, 2016 [in press].

Chapter 3

Low-Noise Operation of Mid-Infrared Quantum Cascade Lasers Using Injection Locking

Hercules Simos, Adonis Bogris and Dimitris Syvridis

Additional information is available at the end of the chapter

http://dx.doi.org/10.5772/66941

Abstract

Quantum cascade lasers are the most promising optical source for emission in the mid-infrared and THz region, and they are already used in a large number of applications such as free-space communications, absorption spectroscopy, sensing and so on. In all these applications, the noise properties of the optical sources are critical for the system performance. In this work, the authors present a theoretical study on the intensity noise characteristics of quantum cascade lasers (QCLs) under external non-coherent optical injection. The injection locking has been proven in the past beneficial for noise properties of bipolar lasers, and thus this technique is utilized here in quantum cascade lasers. With the help of various analytical and numerical models, it is shown that intensity noise reduction can be achieved in the operation of the so-called locked slave laser compared to its free-running values. The detailed analysis reveals the contribution of the various noise sources to the intensity noise of the laser and how they affect the injection locking process. Using different numerical models, two distinct schemes are investigated, analysed and discussed, injection on the lasing mode or on non-lasing residual modes of the slave laser cavity.

Keywords: quantum cascade laser, optical injection locking, noise

1. Introduction

Quantum cascade lasers (QCLs) are a relatively new class of optical sources exploiting radiative intersubband transitions in semiconductor heterostructures, giving operation in a wide range of wavelengths, from 3.5 to 150 μm [1]. After the demonstration of the first QCL in

1994 [2], there has been huge effort for the implementation of sources similar to the ones commonly used in the infrared regime, like single-mode or powerful tunable lasers for either continuous wave or pulsed operation at room temperature (RT). Today, QCL technology is the dominant source for emission in the mid-infrared and THz region. The mid-infrared spectral region hosts a large number of applications including free-space communications, absorption spectroscopy, chemical and biological sensing, and LIDAR (LIght Detection and Ranging) applications. Although QCL technology has reached a certain degree of maturity, many of their physical properties such as the modulation, noise and nonlinear properties are still under investigation and progress, leading to many important contributions in the field. Especially, the noise properties of QCLs are extremely important for spectroscopy and sensing applications as they define the system sensitivity.

Strong effort has been put to identify the specific properties that differentiate these devices from their interband counterparts, the diode lasers, which emit in the near infrared regime [3–7]. In this context, mature techniques from the near infrared wavelength region have been adopted in the mid infrared region to provide insight in the operation and optimize the emission characteristics of QCLs and related systems. One of these techniques, the optical injection locking (OIL) has been investigated extensively for classic (interband) diode lasers in order to exploit its fundamental property of phase synchronization between two independent laser sources [8]. In OIL, a laser source which is considered as "slave" is optically injected by a second "master" laser, and as a result of the locking procedure, the slave laser emits at the wavelength of the master laser and follows its phase fluctuations. Apart from this fundamental property, OIL has been also proposed and utilized in order to demonstrate amplitude noise and phase noise squeezing [9]. In the mid infrared region, the intensity noise of injection-locked QCLs was recently investigated theoretically and experimentally [10, 11], and the obtained results revealed the potential of the locked (slave) laser to operate under reduced intensity noise levels compared to the free-running operation. From the application point of view, absorption spectroscopy would substantially benefit from this reduction of the laser intensity noise, which leads to higher signal-to-noise ratios and thereby enables lower detectable concentrations in corresponding absorption experiments.

In this chapter, the authors analyse the emission characteristics of mid infrared injection-locked QCLs in terms of the intensity noise using theoretical numerical models. The analysis has the following stages. First, a review of the developed computational tools for the proper description of the operation and various aspects of QCLs, such as the noise characteristics and the modulation properties, is carried out. Second, the investigation of the intensity noise characteristics of free-running QCLs, which serves as a calibration/optimization tool for the various numerical models. The third part focuses on the noise performance of injection-locked QCLs and how it is affected by the system parameters. For the injection-locked QCLs, two separate cases are considered, the single-mode scheme (injection in a single-mode slave laser) and the multimode scheme (injection on the main or side modes of a multimode slave laser). Furthermore, special cases for optimized low-noise operation are identified and discussed.

http://dx.doi.org/10.5772/66941

2. Numerical tools

2.1. Rate equation models

The numerical rate equation (RE) models are based on the assumption of the laser homogeneity in the spatial coordinate (direction of propagation) and thus provide a very efficient calculation tool for the dynamics of interband single-mode semiconductor lasers [like Distributed Feedback Lasers (DFB), external cavity lasers, etc]. Lately, RE models have been modified and used to study the emission characteristics of intersubband QCLs. Such a RE model is utilized for the investigation of the intensity noise of single-mode QCLs [12] and presented below.

For the purposes of the noise investigation, the band structure of a QCL is described with a three-level approximation, and it is shown in **Figure 1**. As usually, several assumptions have been made in order to simplify the sophisticated energy structure of the quantum well of each stage. Photons are emitted during a carrier transition from level 3 to level 2, whereas optical phonon emission occurs at the transition from level 2 to level 1. It is assumed that the electrons from the injector of the previous stage are received by energy

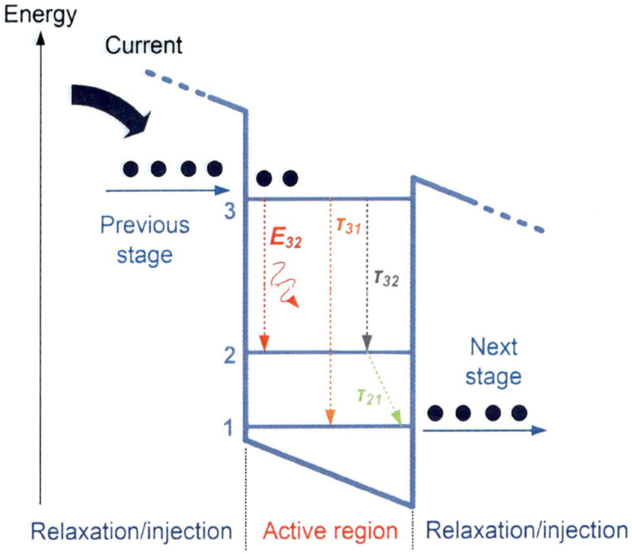

Figure 1. Schematic of the three-level energy band structure of the mid-infrared quantum cascade laser (QCL) considered for the noise investigation.

level 3, and the electrons from level 1 relax to the miniband and injector of the next stage. The radiative and non-radiative transitions from level 3 to levels 2 and 1 are also taken into account. The carrier and photon dynamics are described by rate equations for the carrier number population for each level and the photon number. The RE system assuming single-mode operation reads [12, 13]:

$$\frac{dN_3}{dt} = \frac{\eta I}{q} - \frac{N_3}{\tau_{32}} - \frac{N_3}{\tau_{31}} - g(N_3 - N_2)S + F_3 \tag{1}$$

$$\frac{dN_2}{dt} = \frac{N_3}{\tau_{32}} - \frac{N_2}{\tau_{21}} + g(N_3 - N_2)S + F_2 \tag{2}$$

$$\frac{dN_1}{dt} = \frac{N_3}{\tau_{31}} + \frac{N_2}{\tau_{21}} - \frac{N_1}{\tau_{out}} + F_1 \tag{3}$$

$$\frac{dS}{dt} = Zg(N_3 - N_2)S + Z\beta\frac{N_3}{\tau_e} - \frac{S}{\tau_{ph}} + F_S \tag{4}$$

In Eqs. (1)–(4), $N_i = N_i(t)$ is the carrier number at the energy level i (i = 1, 2, 3), $S = S(t)$ is the photon number, g the gain coefficient, β is the coupling of the spontaneous emission to the optical mode and Z is the number of cascaded stages. The time constants: τ_{ph} is the photon lifetime; τ_e is the carrier recombination time and τ_{ij} the energy-level transition characteristic times. Finally, I is the injection current and η is the current injection efficiency. Also, it is assumed that all electrons which move from the stage to the injector escape to the next stage. The terms $F_i = F_i(t)$ (i = 1, 2, 3, S) represent the noise contribution based on the Langevin force formulation including the various noise correlations; appropriate expressions can be found in Ref. [12]. Taking into account all the non-zero correlations properly, results in a complicated nonlinear system which is difficult to treat numerically. However, analytical calculations of the relative intensity noise (RIN) spectrum based on the RE system (1)–(4) and presented in Ref. [12] identified the separate contribution of each noise term. The results revealed that the dominant contributions to the total RIN come from the carriers in level 3 and the photons, as also confirmed by the experimental results in Ref. [14]. Thus for the usual operating conditions, the cross-correlation terms can be neglected, and the numerical solution of the RE system is simplified. Based on the above assumption, throughout this study, the noise terms F_3, F_2 and F_S have been taken into account assuming the correlations of [12]:

$$\langle F_3 F_3 \rangle = 2\left(g N_3 S + \frac{N_3}{\tau_e}\right) = 2D_{33} \tag{5}$$

$$\langle F_2 F_2 \rangle = 2\left(g N_3 S + \frac{N_3}{\tau_{32}}\right) = 2D_{22} \tag{6}$$

$$\langle F_S F_S \rangle = 2Z\left(g N_3 S + \beta\frac{N_3}{\tau_e}\right) = 2D_{SS} \tag{7}$$

The coefficients D_{ij} represent the correlation strength between the various noise terms. The noise sources F_3, F_2 and F_S in Eqs. (1)–(4) are calculated by

$$F_i = F_i(t) = \sqrt{\frac{2D_{ii}}{dt}}\, x_i \tag{8}$$

In Eq. (8), dt is the integration time step used here to determine the noise bandwidth, and x_i ($i = 3, 2, S$) are independent random Gaussian variables with zero mean value and unit standard deviation.

In order to account for the phase information of the fields which is critical in the injection locking process, the model has been modified: the photon number and the phase have been replaced by the complex field. The time evolution of the field [Eq. (9)] arises from the transformation of Eq. (4) as in Ref. [15]:

$$\frac{dE}{dt} = \frac{1}{2}\left[Zg(N_3 - N_2) - \frac{1}{\tau_{ph}}\right](1 + i\,\alpha_{LEF})E + E_{sp} + \frac{\sqrt{\kappa_{inj}}}{\tau_r} E_{inj} - i\delta\omega E \tag{9}$$

Here $E = E(t)$ is the complex field normalized in terms of photon number ($|E|^2 = S$). The phase change is introduced through the linewidth enhancement factor (α_{LEF}). The spontaneous emission E_{sp} is a complex number and it is given by

$$E_{sp} = \sqrt{\frac{1}{2dt} Z\left(g\,N_3 + \beta\frac{N_3}{\tau_e S}\right)}\,(x + iy) \tag{10}$$

where x and y are random Gaussian variables with zero mean value and unit standard deviation. Equation (10) has been derived following the procedure in Ref. [15].

The two last terms on the r.h.s. of Eq. (9) represent the injection locking process in which a "master" laser is injected into the "slave" laser [16]. E and E_{inj} are the fields of the slave and the master laser, respectively, and their frequency difference is $\delta\omega = 2\pi\cdot df$. The power injection ratio κ_{inj} is defined as the optical power ratio of the master to free-running slave laser (P_{inj}/P). τ_r is the round-trip time of the laser cavity under injection (slave). The system consists of the carrier rate Eqs. (1)–(3) and the noise terms (5)–(7) and (10). In order to simulate a free-running laser, the injection terms in Eq. (9) should be neglected.

2.2. Time domain travelling-wave model

Although the RE model is a very efficient and accurate tool for investigating the majority of applications with semiconductor lasers, there are some cases that require even more sophisticated laser description. In particular, for the noise investigation of QCLs, a modified scheme of injection locking has been proposed and analysed by the authors; in this scheme, the injection of the master laser is carried out in a different than the dominant longitudinal mode of the slave laser. Such a configuration requires the description of the spectral characteristics of the laser, a property which can be treated easily by a travelling-wave (TW) model. The TW models can describe the multimode spectrum of a Fabry-Perot cavity and take into account the spectral profile of the material gain, both being necessary conditions in order to simulate the case of optical injection into secondary longitudinal modes. Furthermore, the TW models take into account the spatial distribution of the carriers and the field along the propagation direction, and thus they offer a more accurate description of the carrier dynamics in the gain medium than the RE models.

In the TW model, the carrier dynamics in the three-level QCL approximation shown in **Figure 1** is assumed to depend on the spatial coordinate z and is described by a set of two rate equations [12, 13] for the carrier densities N_3 and N_2 [since no carrier loss between stages is assumed, the system is independent of N_1 and Eq. (3) may be neglected]:

$$\frac{dN_3}{dt} = \frac{\eta I}{q} - \frac{N_3}{\tau_{32}} - \frac{N_3}{\tau_{31}} - R_{st}(z,t) + F_3 \tag{11}$$

$$\frac{dN_2}{dt} = \frac{N_3}{\tau_{32}} - \frac{N_2}{\tau_{21}} + R_{st}(z,t) + F_2 \tag{12}$$

Here, the carrier numbers depend on the spatial coordinate z; thus, $N_3 = N_3(z,t)$ and $N_2 = N_2(z,t)$ are the local carrier numbers of the corresponding levels 3 and 2 between which the laser transition takes place; τ_{ij} is the characteristic transition time. The terms F_3 and F_2 represent the carrier noise, which follows the Langevin force formulation like in the RE model [Eqs. (5) and (6)], properly modified to account for the spatial dependence. The field propagation inside the active region is described by the travelling-wave equation for the forward and the backward slowly varying envelopes $E^{\pm}$ [11, 17]:

$$\frac{dE^{\pm}(z,t)}{dz} = \frac{1}{2} Zg(z,t)\left[\gamma e^{-\gamma t} \otimes E^{\pm}(z,t)\right] - i\frac{1}{2} Zg(z,t)\,\alpha_{LEF}\, E^{\pm}(z,t) - \frac{1}{2} a_l\, E^{\pm}(z,t)$$
$$+ \kappa_{DFB}\, E^{\mp}(z,t) + E_{sp}(z,t) \tag{13}$$

$E^{\pm}$ are the forward (+) and backward (−) propagating fields, g is the gain coefficient, Z is the number of cascaded gain stages, α_{LEF} is the linewidth enhancement factor and a_l is the total linear loss. g is the gain per unit length (m^{-1}) and depends on the carrier number of the levels 3 and 2:

$$g(z,t) = \frac{\Gamma g_0}{v_g}\left[N_3(z,t) - N_2(z,t)\right] \tag{14}$$

In Eq. (14), Γ is the confinement factor, v_g is the group velocity in the active region and g_0 is the differential gain constant in s^{-1}. The stimulated emission term in Eqs. (11) and (12) is given by

$$R_{st}(z,t) = \frac{g(z,t)V}{\hbar\omega wd} \sum_{i=\pm} \Re\left\{E^{i*}(z,t)\left[\gamma e^{-\gamma t} \otimes E^{i}(z,t)\right]\right\} \tag{15}$$

The incorporation of a material gain profile with a finite spectral width in the travelling-wave equations is carried out by a filtering process in the time domain. Here, the gain profile is approximated with a Lorentzian function. As shown in Eqs. (13) and (15), the filtering process is described by the convolution product (symbol ⊗) of the field with the Lorentzian function of width γ. The operator $\Re()$ is the real part of a complex number. The simulation of multi-mode or single-mode operation is accomplished by adjusting the coefficient κ_{DFB} in Eq. (13), which represents the coupling between the forward and backward propagating waves due to the DFB grating in the waveguide. The term $E_{sp}(z,t)$ is the spontaneous emission noise per unit length; its expression is derived using the Langevin formulation of Ref. [10], and it is based on the spontaneous emission factor β (e.g. the coupling of the spontaneous emission to the lasing mode). Other parameters appear in **Table 1**.

The whole active region of length L is sliced into N sections of length dz each and Z stages with the same structure. The cavity effect and the optical injection of the master laser are modelled

Parameter	Symbol	Value
Active region length	L	2 mm
Active region width	w	8 μm
Active region depth	d	2 μm
Number of stages	Z	25
Facet power reflectivities	R_1, R_2	[0.27, 0.27]
Photon lifetime	τ_p	4.3 ps
Waveguide loss	a_i	1200 m^{-1}
Current injection efficiency	η	0.45
Gain coefficient	g	1.4×10^4 s^{-1}
Gain spectral width (FWHM)	γ	7 meV
Spontaneous emission factor	β	10^{-6}
Linewidth enhancement factor	α_{LEF}	0.5
Refractive index	n	3.2
Emission wavelength	λ	5 μm
Phonon scattering time $E_3 \rightarrow E_2$	τ_{32}	2.1 ps
Phonon scattering time $E_3 \rightarrow E_1$	τ_{31}	2.6 ps
Phonon scattering time $E_2 \rightarrow E_1$	τ_{21}	0.3 ps
DFB coupling coefficient	κ_{DFB}	10 m^{-1}
Injection coefficient (power)	κ_{inj}	Variable

Table 1. Physical and structural parameters used for the simulations of QCLs.

with boundary conditions. The solution of Eq. (13) is carried out in the time domain using an iterative algorithm to give the output field $E^{+}(L,t)$.

2.3. Noise calculations

The relative intensity noise (RIN) spectrum can be obtained from the output power fluctuations, and it is defined by

$$RIN(f) = \delta P\,(f)^2 / \langle P \rangle^2 \tag{16}$$

where P is the laser output power, f is the frequency, $\delta P(f)$ is the power fluctuations and the symbol $\langle \cdot \rangle$ represents the mean value. In the experiments, the RIN spectrum is calculated through Eq. (16) with the help of an electrical spectrum analyser (ESA). In the simulations, the instantaneous power fluctuations can be calculated from the output power of the laser in the time domain over a long time series with span T, as $\delta P(t) = P(t) - \langle P(t) \rangle$. The RIN spectrum then reads

$$RIN(\omega) = \frac{1}{\langle P(t)\rangle^2} \frac{1}{T} \left| \int_0^T \delta P(t)\, e^{-j\omega t}\, dt \right|^2 \tag{17}$$

Usually, the RIN value is calculated by averaging a narrow bandwidth at the low-frequency range where the spectrum is flat. Throughout this investigation, a single frequency window around 100 MHz has been used.

3. Intensity noise of free-running QCLs

The first step of the investigation regarding the intensity noise of injection-locked QCLs was to simulate the properties of a free-running QCL in order to confirm the validity of the RE model and the assumptions made for the contribution of the noise sources. Thus the RE model was used to simulate a single-mode QCL and compare the results with the corresponding results generated with the analytical theory of Gensty and Elsasser [12]. For this reason, all the parameters are taken from the work of Gensty and Elsasser [12] and shown in **Table 1**. **Figure 2** shows the RIN of the free-running QCL versus the output optical power (left) and the intensity noise spectrum in two cases (right). On the left side of **Figure 2**, the line corresponds to the analytical results generated with the theoretical model of Gensty and Elsasser [12], whereas the squares correspond to the numerical results generated by solving RE model under the same conditions. Both the analytical and numerical values of the RIN versus the optical power were obtained in the low-frequency range, and in the case of the RE model, a window of 150 MHz has been averaged. In order to reduce the statistical uncertainty due to the noise sources, the RIN value for each power level (current) is calculated by the average of 25 runs of the RE system with independent noise realizations.

As shown in **Figure 2**, the numerical results (squares) are in perfect agreement with the analytical ones (line), despite the simplifications made for the noise sources. This confirms the assumption that the dominant contributions to the total RIN come from the carriers in level 3 and the photons. The calculated intensity noise spectra for two different cases of output

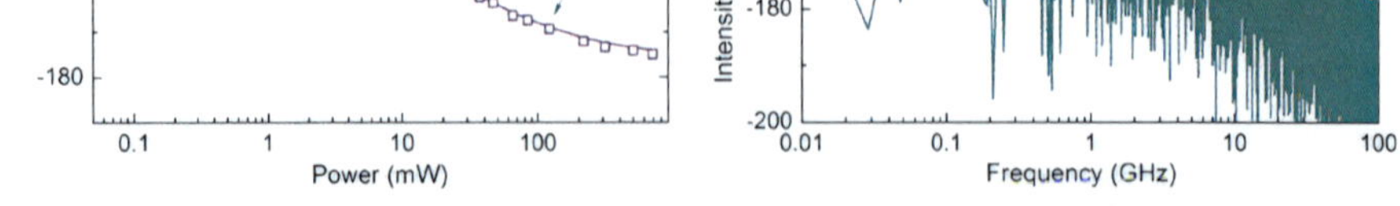

Figure 2. Left: The RIN of a free-running QCL versus the output optical power calculated with analytical (line) and numerical model (squares). Right: The intensity noise spectrum for two cases (points A and B of the RIN versus power curve).

optical power are also shown on the left of **Figure 2**. For low to moderate output power, the RIN spectrum is flat in the low-frequency regime (<1 GHz), and it fades out with frequency as shown by the spectrum of case A (top). In all cases, the spectrum in the low-frequency range and up to a few GHz is flat. Furthermore, the low-frequency plateau decreases with power, and the spectrum is flat up to a few tens of GHz as shown by the spectrum, which corresponds to case B (bottom). The flat low-frequency region and the disappearance of the resonance in the intensity noise spectrum are characteristics of damped carrier and thus photon relaxation oscillations. This behaviour is typical in QCLs in which the relaxation of the carriers is dominated by an ultrafast mechanism, the phonon scattering. This property is particular interesting for direct modulation and related applications which are modulation bandwidth demanding.

4. Intensity noise of injection-locked QCLs A: single-mode scheme

In this paragraph, the authors report the results of the investigation regarding the intensity noise properties in the most generic scheme of the injection locking method. In this scheme, the master and the slave QCLs are assumed to be ideal and emit in a single longitudinal mode; thus, the RE model was used for the simulations of the injection locking. The investigation is based on the effect of the two most important parameters of the injection locking process, the injection strength and the frequency detuning between the master and the slave laser.

4.1. Effect of the injection strength and the frequency detuning

By definition, the RIN properties of a laser are directly related to the emitted optical power, so it is reasonable to investigate first the injection process in terms of the output power. Furthermore, apart from the RIN, the role of the injected power from the master laser into the slave laser is significant on the locking process itself. **Figure 3a** illustrates the emitted optical power of the slave laser versus the current for different values of the injected optical power from the master laser (P_m), assuming the simplest case of zero master-slave frequency detuning (df = 0). A significant increase of the output power is observed below and right after the threshold current in all cases of injection compared to the free-running slave QCL case (P_m = 0). In the case of strong injection, the lasing threshold of the slave laser vanishes because of high injected power which turns on the lasing action in the slave laser, even if the current is very low. Essentially, the master laser acts as an optical pump for the slave laser, as it induces a huge increase of the photon number in the slave cavity. Above and close to the free-running threshold, the optical power of the slave laser gets even higher as the injected optical power increases. Furthermore, the power increase is practically minimized when moving away from the free-running threshold. Here, the emitted power is comparable with the power of the free-running slave laser.

Figure 3b illustrates the RIN levels of the slave laser over the entire range of the current corresponding to the conditions of **Figure 3a**. However, the RIN is plotted versus the emitted optical power of the slave laser instead of the current, in order to obtain a comparison of the RIN perfor-

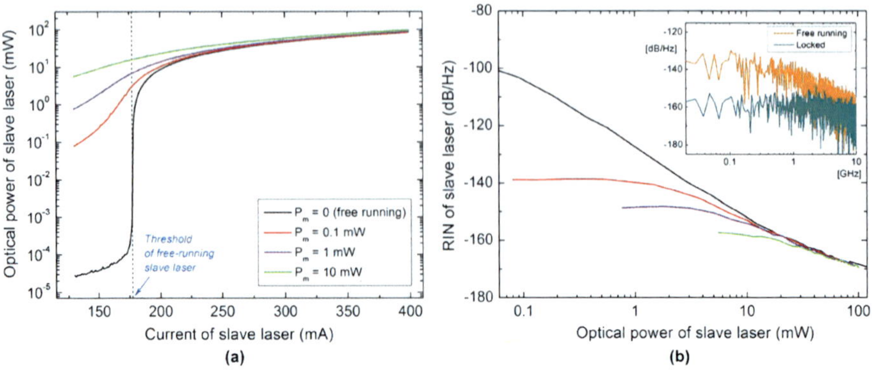

Figure 3. (a) The emitted optical power from the slave laser versus the injection current for different levels of the injected power from the master laser (P_m). (b) The RIN of the slave laser versus the emitted optical power. Insets show noise spectra of free-running and locked slave laser. In all cases, it is assumed zero frequency detuning between the master and slave lasers.

mance for different optical injections from the master laser but at the same output power levels. Additionally, a procedure for taking into account the right amounts of vacuum noise has been adopted for the calculations. The master QCL was driven at a high current to give the minimum possible RIN value of −175 dB/Hz, and its optical power was attenuated to the required values for injection to the slave laser. During every attenuation process, the degradation of the signal has been taken into account by introducing the vacuum noise properly. The resulting sets of (P_m, RIN) are (10 mW, −170 dB/Hz), (1 mW, −161 dB/Hz) and (0.1 mW, −151 dB/Hz).

As shown in **Figure 3b**, the RIN of the slave laser is reduced dramatically compared to the free-running values in all cases of injection. For low optical power of the slave laser (below and around the threshold current), the RIN is flat and is mainly determined by the injected power of the master laser and its intrinsic noise levels. The RIN gradually decreases with the output optical power, and up to a few mWs, it is still much lower than the corresponding free-running value. For current values far above threshold, the RIN reduction is saturated and similar to the free-running values are obtained. The inset of **Figure 3b** illustrates the intensity noise spectrum of the slave laser, in the case of free-running operation (orange line) and in the case of locked operation (cyan line), considering an injection power ratio (κ_{inj}) equal to 1 (1.7 mW). A suppression of ~20 dB on the low-frequency RIN of the locked slave laser is observed following the predictions of **Figure 3a**. However, as previously discussed, the actual RIN suppression is lower if compared to the same optical power, since the locking process results in a large power increase. Thus, in order to identify the actual noise performance, the RIN is always referenced/plotted at emitted power and not the current.

The RIN characteristics of the injection-locked QCLs can be explained with the help of the analytical calculations regarding the contribution of the different noise sources to the total RIN. As explained in Ref. [10], at the low output power regime (up to 5 mW in our case), the spontaneous emission of a free-running QCL (term R_{SS}) is unsaturated and dominant over the carrier noise (terms R_{33} and R_{22}). When photons are injected from the master laser, the spontaneous emission noise is suppressed significantly similar to what happens in bipolar semiconductor lasers; furthermore, since it is the dominant noise source, the total RIN is highly reduced compared to the free-running value. The situation is different in the high-power regime (far from the laser threshold) where the dominant contribution to the laser intensity noise comes from the carrier noise (R_{33}). Here, both the spontaneous emission noise and the carrier noise are already strongly saturated/suppressed. Therefore, in this case, the injection locking has very little or no effect on the RIN of the slave QCL. The domination of carrier noise in this regime imposes a fundamental limit on the potential RIN suppression achieved with injection locking directly on the oscillating mode of the QCL.

The frequency difference (detuning, df) between the master and the slave lasers is the other parameter that affects significantly the injection locking process. In detail, although the locking process can occur within a finite frequency detuning range between the two lasers, there is a dependence of the emitted power on the detuning which defines the so-called locking bandwidth of the process. Thus the locking bandwidth establishes the range of frequency difference between the two lasers in which the locking process is possible. In the case of QCLs, the locking bandwidth has been calculated analytically in previous works [7, 18] but with no correlation to noise properties. Here, the locking bandwidth is connected to the intensity noise of the emitted field from the slave QCL for various parameters. For this reason, in the calculations, an effective bandwidth was used, defined as the bandwidth of the noise reduction.

Figure 4a shows calculations of the optical power (red line) and the low-frequency RIN (violet circles) of the slave laser versus the frequency detuning between the master and slave lasers (df). The RIN distribution around the lasing mode is mapped with a large number of independent simulation runs in the investigated range (from −15 GHz to 15 GHz) with respect to the lasing frequency (0 GHz). For the calculations on the top of **Figure 4a**, the power of the free-running slave laser was P_{fr} = 10 mW (dashed black line) giving a RIN value of RIN_{fr} = −152 dB/Hz (the shadowed regions correspond to operation with RIN values below the free-running value). The injected power from the master laser was P_m = 1 mW (top left) and P_m = 10 mW (top

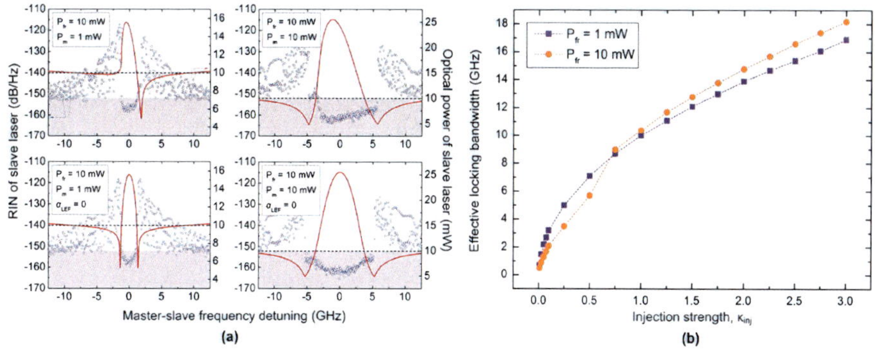

Figure 4. (a) The locking maps for two cases of injected power and (b) the effective locking bandwidth of the injection locking process.

right) having RIN values RIN_m = −161 dB/Hz and RIN_m = −170 dB/Hz, respectively. In both cases of injection, the bandwidth of the locking process can be extracted from the increased output power of the slave laser around the zero detuning. When low injection ratio is considered (κ_{inj} = 0.1, P_m = 1 mW), the locking occurs for ~2 GHz of frequency detuning and a reduction of the slave laser RIN in the order of 3–5 dB from the free-running value is observed. For frequencies outside the locking regime, the slave laser seems to operate in a "modulation" regime [19]. In this type of operation, the power is modulated as observed on the time traces and the corresponding RF spectra which exhibit peaks in the corresponding random frequencies. Such regimes most likely correspond to unstable locking, unlocked or chaotic oscillations, and they have been recently investigated in Ref. [20]. The asymmetry and the exact structure of the obtained locking map with respect to the difference in the negative and positive detuning are related to the non-zero linewidth enhancement factor [16]. Calculations under the same conditions but with a zero linewidth enhancement factor (α_{LEF} = 0) confirm this claim as the locking maps are perfectly symmetrical (bottom row of **Figure 4a**). When considering a higher injection ratio (κ_{inj} = 1, P_m = 10 mW), apparently the locking range is wider (~7 GHz) and the slave laser RIN is further reduced about ~10 dB with respect to the free-running case. Following the trend, the optical power of the slave laser is significantly increased compared to the case with κ_{inj} = 0.1.

Next, the effective locking bandwidth (defined as the bandwidth in which the slave QCL operates with lower noise than the free-running value) was calculated versus the injection strength, for different driving conditions of the slave QCL. Furthermore, in order to avoid the asymmetry of the locking region in the calculation of the effective bandwidth, the linewidth enhancement factor was set to zero (α_{LEF} = 0). The calculations have been carried out for P_{fr} = 1 mW and P_{fr} = 10 mW, and the results are plotted versus κ_{inj} in **Figure 4b**. In both cases, the effective locking bandwidth increases rapidly with the injections' strength. For low power of the slave QCL (P_{fr} = 1 mW), the injection locking gains more bandwidth in the low injection strength regime (κ_{inj} < 1) compared to the case of P_{fr} = 10 mW. In the high injection strength regime (κ_{inj} > 1), this behaviour is reversed since the locking process saturates earlier when the slave power is driven with low current.

4.2. Optimized noise operation in the single-mode injection locking scheme

With a second view to the locking maps of **Figure 4a**, a remarkable situation is revealed. The RIN of the slave laser within the locking range remains almost constant, whereas at the same time, the optical power is reduced significantly when moving towards the boundary regions. For example, in the case of P_{fr} = 10 mW (top right), close to the zero detuning (df = 0), the emitted power is 25 mW, and for detuning of 4 GHz, the emitted power is 8 mW. While the ratio in the optical power is 5 dB, the RIN changes only 2 dB, which indicates that there is operation with lower RIN under the same power levels. This observation leads to an important evidence: injection applied near the edges of the locking regime may drive the slave laser to operate under further reduced intensity noise due to the lower power. The actual RIN suppression which results from injection locking assuming the above conditions can be obtained with focusing on different locking regions with specific frequency detuning between the mas-

http://dx.doi.org/10.5772/66941

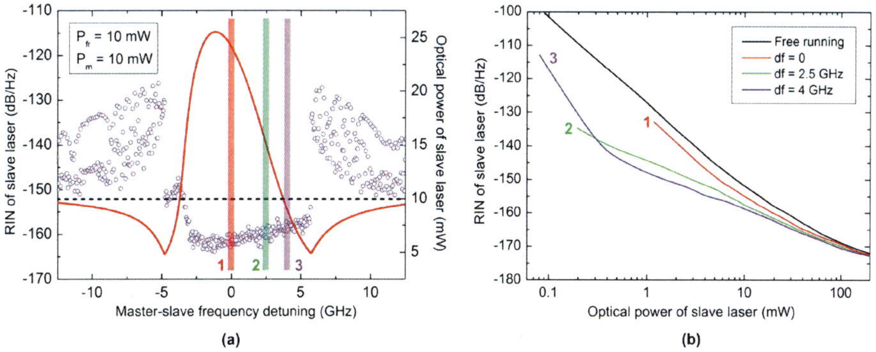

Figure 5. (a) The locking map indicating different regions with respect to the frequency difference between the master and the slave QCL and (b) the RIN of the slave laser under no injection and under injection locking at the frequency detuning values highlighted in (a).

ter and slave QCLs (df). To this end, the RIN of the slave laser was calculated versus the emitted optical power from the slave laser assuming a constant injection ratio instead of constant injected power in order to obtain RIN values of the slave laser in the low-power regime.

Figure 5 illustrates the results for $\kappa_{inj} = 1$ (the mean value of 25 independent runs). In the conventional case of injection locking directly on the lasing mode (df = 0), the RIN of the slave laser is constantly lower 4 dB than the free-running value up to 10 mW, however gradually approaching the free-running value, for even higher power (**Figure 5b**, red line). For injection in the second regime indicated in **Figure 5a** (df = 2.5 GHz), even lower RIN levels are observed before the saturation (**Figure 5b**, green line). In our analysis, the highest possible RIN suppression was found towards the positive edge of the locking regime (df = 4 GHz), where the RIN value was lower 21, 7 and 3 dB than the free-running value at 1, 10 and 50 mW of emitted power, respectively (**Figure 5b**, purple line).

5. Intensity noise of injection-locked QCLs B: multimode laser scheme

In the most common scheme of injection locking, the injection is realized in such a way that the wavelength of the master laser approximately matches the wavelength of the single-mode slave laser, e.g. injection into the lasing main longitudinal mode of a slave DFB laser. This technique has been adopted for the investigation of the previous paragraph. In addition to this scheme, several variations of the injection locking technique have revealed interesting characteristics in the case of interband diode lasers. For example, sub-shot noise performance (amplitude squeezing) has been achieved when optical injection is applied into a secondary longitudinal mode [21]. In order to optimize their properties, this method was applied to QCLs, and investigation of the noise properties and the locking potential has been carried out, as analysed in the next paragraphs.

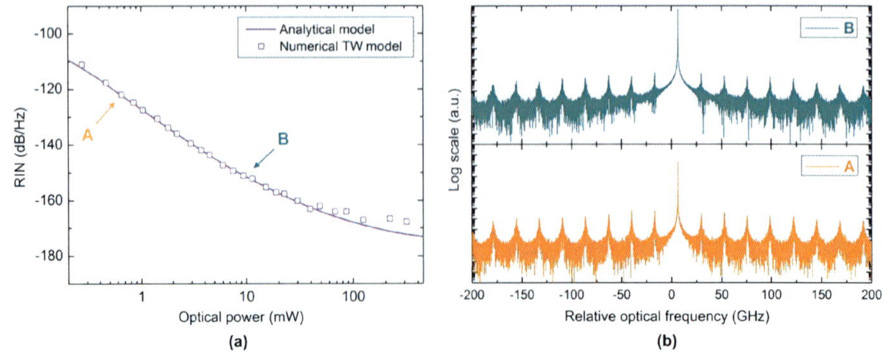

Figure 6. The RIN (a) and the optical spectrum (b) of the free-running QCL simulated with the TW model.

First, the TW model was used to reproduce the operation of a free-running QCL. **Figure 6a** illustrates the RIN of the free-running QCL simulated with both the TW and the analytical models with the parameters of the previous paragraph. The TW model is assumed to operate as a single-mode DFB laser, adjusting properly the grating coupling coefficient (κ_{DFB}). It is obvious that although the TW model is conceptually different from the RE model due to its inherent multi-longitudinal mode spectrum and the gain spectral profile, it reproduces fairly the noise characteristics of the single-mode QCL. As shown in **Figure 6b**, which illustrates the optical spectra in the two cases marked in **Figure 6a**, the side mode suppression ratio is very high even close to the threshold (spectrum on the bottom of **Figure 6b**).

Then, the TW model was used to investigate the noise properties of injection-locked QCLs with respect to the frequency detuning df. For this set of simulations, the power of the slave laser was set at 10 mW, and two cases of injection strength were considered: the master laser was driven at 1 and 10 mW (κ_{inj} = 0.1 and κ_{inj} = 1, respectively). In the case of injecting directly into the lasing mode (**Figure 7**, mode 0), the qualitative behaviour of the RIN with respect to the frequency detuning df is similar to the one generated with the RE model and analysed in the previous paragraphs. Here, the power and RIN profiles are asymmetrical with respect to the master-slave detuning, but also with respect to the zero frequency detuning due to the non-zero linewidth enhancement factor of the QCL. With higher injection the slave QCL exhibits higher locking bandwidth and optical power as well as lower noise in the locking range. Apart from this expected behaviour, the case of injection into a non-lasing secondary side mode was investigated. **Figure 7** demonstrates that the optical injection mechanism in a secondary mode (modes −1, +1) potentially leads to locking of the slave laser to the master wavelength and with similar characteristics (RIN, power), as in the case of injecting directly in the lasing mode. This is reasonable, since the adjacent to the lasing cavity modes is very close to the peak of the gain spectrum of the slave QCL, and the modes experience almost the same gain. The characteristics of injection on the main and the side modes have also been investigated experimentally and numerically by the authors in Ref. [11], and the results confirmed the

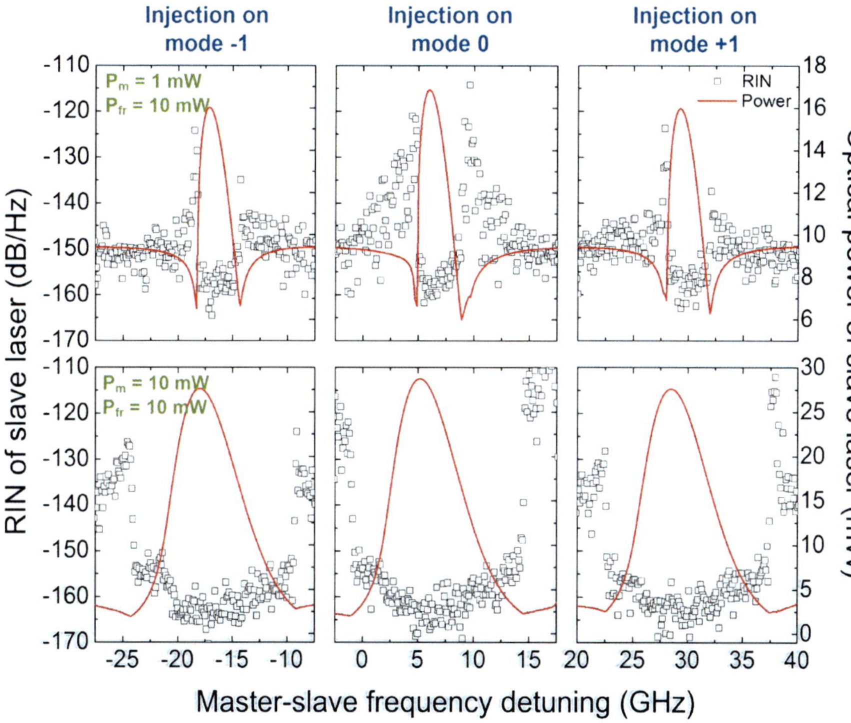

Figure 7. RIN and optical power of the slave QCL versus the frequency detuning for injection on the lasing (mode 0) and on the two adjacent longitudinal modes of the cavity (mode −1, +1).

results of this work. This is very important from the applications' point of view, since in practical conditions, it is not always easy to have QCLs (master-slave) emitting at the same wavelengths.

A more interesting and fundamentally different case occurs when the injection is realized into a resonant mode away from the gain peak of the slave QCL. This case has been simulated similarly to the previous cases but now for an extended master-slave frequency difference range of up to ±600 GHz in order to cover many of the modes within and outside the gain bandwidth. For this set of calculations, the power of the master and the same laser was the same, $P_{fr} = P_{ma} = 10$ mW. In order to give a better view, the RIN and the optical power of each of the simulated modes of the slave QCL have been plotted in **Figure 8** in a relative order showing their position within the optical spectrum and the gain profile of the laser. The red dashed line and the shadowed area denote the free-running value of the optical power (10 mW) and operation with lower RIN than the free running value (−152 dB/Hz), respectively.

The optical power of the slave QCL and the locking bandwidth fade slowly with the number of mode on which injection is carried out, due to the lower material gain available away

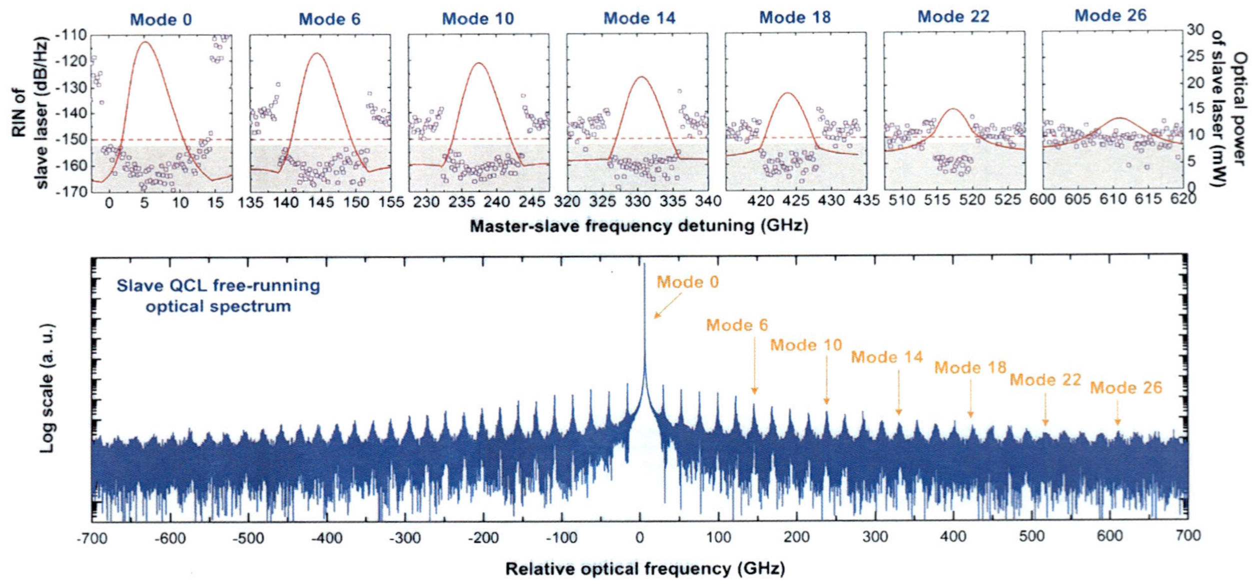

Figure 8. The RIN and the optical power versus the frequency detuning in the case of injection on different modes of the slave QCL.

of the gain peak. This characteristic is more intense after the 14th mode, as seen in **Figure 8**. Furthermore, for the same reason, the asymmetry of the locking characteristics with respect to the central frequency of each mode vanishes towards the edges of the spectrum. On the other hand, the RIN within the locking bandwidth is almost constant for many modes, as long as the injection locking takes place close to the peak of the gain spectrum of the slave laser. In detail, the lower average RIN within the locking bandwidth for all the modes up to the 10th is approximately −162 dB/Hz, whereas at the same time, the power decreases. Injection within the locking range of the above modes would result in a RIN suppression of ~10 dB compared to the free-running value. This finding provides us with a significant tool for optimization of the noise suppression, because similar RIN levels are achieved at lower optical power, which results in even higher actual RIN suppression (comparing RIN at the same optical power). This alternative type of injection locking provides stronger noise reduction as long as locking is applied on side modes which lie well within the gain bandwidth of the slave laser. This behaviour is similar to the one observed in the single-mode injection scheme (investigated with the RE model) where the RIN is much lower when the master laser is injected with a small detuning from the frequency of the slave laser (paragraph in Section 4.2). The same characteristics were observed for negative master-slave detuning.

In the case of injection outside of the gain bandwidth, although injection locking may fundamentally occur, the locking bandwidth is drastically reduced and the reduction potential of RIN is minimal. An indicative result is shown in **Figure 8** (mode +26).

6. Conclusions

According to the results reported in the previous paragraphs, for specific driving conditions, the injection locking has the ability to improve the RIN performance of a QCL laser. In summary, optical injection drives the slave laser to operate under simultaneous increased optical power and reduced RIN within the locking range, due to the suppression of the spontaneous emission noise and an increase of the photon number. This behaviour is dominant close to the threshold where spontaneous emission and phonon emission are the dominant noise factors. Far away from the threshold, the carrier noise from energy level 3 becomes stronger, and the total RIN suppression is lower, practically insignificant. Optimized low-noise operation was found with locking under frequency detuned injection between master and slave QCLs, resulting in significant actual RIN suppression (same RIN at lower emitted power) in the unsaturated regime (close to threshold). Furthermore, with an alternative scheme of injection locking on secondary modes of the slave QCL (not the free-running oscillation mode), it was observed that the RIN characteristics resemble the corresponding ones for injection on the main mode and may result in even further enhanced performance due to lower optical power. Based on the above investigation and the corresponding results, the injection locking process has been proven to be an efficient and practical technique to provide low-noise optical sources in the mid-IR region using quantum cascade lasers, which is critical for the majority of the sensing and spectroscopy applications.

Author details

Hercules Simos[1]*, Adonis Bogris[2] and Dimitris Syvridis[3]

*Address all correspondence to: simos@di.uoa.gr

1 Department of Electronics Engineering, Technological Educational Institute of Piraeus, Athens, Greece

2 Department of Informatics, Technological Educational Institute of Athens, Athens, Greece

3 Department of Informatics and Telecommunications, National and Kapodistrian University of Athens, Athens, Greece

References

[1] F. Capasso. High-performance midinfrared quantum cascade lasers. Optical Engineering. 2010;**49**(11):111102.

[2] J. Faist, F. Capasso, D. L. Sivco, C. Sirtori, A. L. Hutchinson, A. Y. Cho. Quantum cascade laser. Science. 1994;**264**(5158):553–556.

[3] S. Bartalini, S. Borri, P. Cancio, A. Castrillo, I. Galli, G. Giusfredi, D. Mazzotti, L. Gianfrani, P. De Natale. Observing the intrinsic linewidth of a quantum-cascade laser: beyond the Schawlow-Townes limit. Physical Review Letters. 2010;**104**(8):083904.

[4] F. P. Mezzapesa, L. L. Columbo, M. Brambilla, M. Dabbicco, S. Borri, M. S. Vitiello, H. E. Beere, D. A. Ritchie, G. Scamarcio. Intrinsic stability of quantum cascade lasers against optical feedback. Optics Express. 2013;**21**(11):13748–13757.

[5] T. Yang, G. Chen, C. Tian, R. Martini. Optical modulation of quantum cascade laser with optimized excitation wavelength. Optics Letters. 2013;**38**(8):1200–1202.

[6] A. Calvar, M. I. Amanti, M. Renaudat St-Jean, S. Barbieri, A. Bismuto, E. Gini, M. Beck, J. Faist, C. Sirtori. High frequency modulation of mid-infrared quantum cascade lasers embedded into microstrip line. Applied Physics Letters. 2013;**102**(18):181114.

[7] C. Wang, F. Grillot, V. Kovanis, J. Even. Nonlinear dynamics and modulation properties of optically injected quantum cascade lasers. In: CLEO/Europe - IQEC 2013; Munich, Germany. 2013.

[8] R. T. Ramos. Optical injection locking and phase-lock loop combined systems. Optics Letters. 1994;**19**(1):4–6.

[9] Y. Lai, H. A. Haus, Y. Yamamoto. Squeezed vacuum from amplitude squeezed states. Optics Letters. 1991;**16**(19):1517–1519.

http://dx.doi.org/10.5772/66941

[10] H. Simos, A. Bogris, D. Syvridis, W. Elsasser. Intensity noise properties of mid-infrared injection locked quantum cascade lasers: I. Modeling. IEEE Journal of Quantum Electronics. 2014;**50**(2):98–105.

[11] C. Juretzka, H. Simos, A. Bogris, D. Syvridis, W. Elsaber, M. Carras. Intensity noise properties of mid-infrared injection locked quantum cascade lasers: II. Experiments. IEEE Journal of Quantum Electronics. 2015;**51**(1):2300208.

[12] T. Gensty, W. Elsasser. Semiclassical model for the relative intensity noise of intersubband quantum cascade lasers. Optics Communications. 2005;**256**(1–3):171–183.

[13] F. Rana, R. J. Ram. Current noise and photon noise in quantum cascade lasers. Physical Review B. 2002;**65**(12):125313.

[14] T. Gensty, W. Elsasser, C. Mann. Intensity noise properties of quantum cascade lasers. Optics Express. 2005;**13**(6):2032–2039.

[15] N. Schunk, K. Petermann. Noise analysis of injection-locked semiconductor injection lasers. Noise analysis of injection-locked semiconductor injection lasers. IEEE Journal of Quantum Electronics. 1986;**22**(5):642–650.

[16] E. K. Lau, H.- K. Sung, M. C. Wu. Frequency response enhancement of optical injection-locked lasers. IEEE Journal of Quantum Electronics. 2008;**44**(1):90–99.

[17] M. Gioannini, M. Rossetti. Time-domain traveling wave model of quantum dot DFB lasers. IEEE Journal of Selected Topics in Quantum Electronics. 2011;**17**(5):1318–1326.

[18] B. Meng, Q. J. Wang. Theoretical investigation of injection-locked high modulation bandwidth quantum cascade lasers. Optics Express. 2012;**20**(2):1450–1464.

[19] I. Petitbon, P. Gallion, G. Debarge, C. Chabran. Locking bandwidth and relaxation oscillations of an injection-locked semiconductor laser. IEEE Journal of Quantum Electronics. 1988;**24**(2):148–154.

[20] T. Erneux, V. Kovanis, A. Gavrielides. Nonlinear dynamics of an injected quantum cascade laser. Physical Review E. 2013;**88**(3):032907.

[21] A. Furusawa. Amplitude squeezing of a semiconductor laser with light injection. Optics Letters. 1996;**21**(24):2014–2016.

Section 2

Applications

Chapter 4

THz QCLs Design Toward Real Applications

Tsung-Tse Lin

Additional information is available at the end of the chapter

http://dx.doi.org/10.5772/65351

Abstract

For highly desired THz applications, we discuss the design and fabrication of THz quantum cascade lasers (QCLs) toward the high temperature and large average output power operations for the real applications with the relatively compact portable size cryogenic cooling systems. We describe the temperature performance parameters of THz QCLs and introduce the recent results of an indirect injection design scheme in the THz region and modulation height active structure design with different barriers and wells for the further design direction. The recent fabricated THz QCLs are combined with the liquid nitrogen (LN) cooling Dewar condenser to demonstrate the relatively compact THz source unit by QCLs. The different injection schemes in THz and barriers-wells height design in the active region introduce one of the directions for the further high temperature and large output power operation of THz QCLs. The relatively compact size THz source unit with a cryogenic system demonstrates the THz QCLs for real applications with the milliwatt order average output operation near liquid nitrogen temperature.

Keywords: Terahertz, quantum cascade laser, semiconductor THz source

1. Introduction

The terahertz (THz) region in the electromagnetic spectrum has drawn much attention due to its wide range of applications in various fields such as spectroscopy, imaging, remote sensing, and communications. Compact THz semiconductor sources are also extremely promising for use in future high-speed and large-capacity local telecommunications applications, especially for those applications operating in the range from sub-THz to a few THz (0.2–2 THz) [1]. The output power of conventional mature radio frequency (RF) electronic devices reduces by 4 orders of magnitude with frequency, which is close to 1 THz in the order of a few microwatts (μW).

High-output-power continuous-wave (CW) operation of optical semiconductor devices is very attractive for overcoming this problem. Quantum cascade lasers (QCLs) [2] are compact semiconductor light sources that utilize carrier recycling and intersubband transitions in repeating quantum well (QW) structures, and they have been demonstrated to operate successfully in the mid-infrared (mid-IR) [2] and THz [3] regions. They are also arguably the only THz solid-state sources with average optical output power levels much greater than one milliwatt (mW). This property of high optical output power with a narrow emission line width is quite attractive for a wide range of THz applications.

The current status of THz QCLs that operate without an external magnetic field has been reported in the spectral range from 1.2 to 5.2 THz [4, 5] with a maximum output power of around 1.01 W in pulsed mode [6]. In contrast to the room temperature operation of mid-IR QCLs, the maximum operating temperature (T_{max}) of THz QCLs is 199.5 K [7] at 3.2 THz. This performance was realized by an optimized state-of-the-art $Al_{0.15}Ga_{0.85}As/GaAs$ structure utilizing longitudinal optical (LO) phonon depopulation for the extraction scheme with resonant tunneling (RT) injection and diagonal emission. However, even this temperature performance still limits their practical use in some applications. Higher operating temperatures for THz QCLs with wider frequency that can at least function at temperatures attainable using Peltier coolers (230–250 K) have been one of the most important topics in recent research. The commonly reported high-temperature performance of THz QCLs roughly follows an empirical limitation depending on their operating frequency, such that $T_{max} \sim h\omega/k_B$. Compared with the near 200 K operation of THz QCLs at 3–4 THz, at low frequencies (<2 THz), THz QCLs are expected to exhibit poorer temperature performance due to this limitation when the frequency decreases. THz QCLs that operate at low frequencies suffer from design difficulties due to the narrow radiative energy separation between the two lasing-subband levels, which is related to the narrow dynamic range of the current density of the laser. Large thermal perturbation effects within the narrow energy level spaces have prevented the expansion of THz QCLs to low-frequency high-temperature operation. This reduction in the dynamic range of the current density with operating frequency can be qualitatively explained by the dependence of RT injection scheme THz QCLs on $k_B T_{max} \sim h\omega$ [8]. It is known to be difficult to achieve both low-frequency lasing and high-temperature operation.

On the basis of some successful solutions in mid-infrared QCLs, the different barrier height shows some solution to improve the device performance for different directions. For example, utilizing the height barriers to reduce the high temperature parasitic leakage currents [9, 10] and step wells for improving internal quantum efficiencies [11]. Furthermore, with the high technique of crystal growth, the optimization of the individual barrier and well's height at the suitable place is possible to give the solution and one more design freedom of recent stagnated structure design of THz QCLs. Considering these research and our previous studies on high Al composition $Al_xGa_{1-x}As/GaAs$ design [9], it indicates one of the possible directions for combining the high Al composition structure with variable well-barrier height design [12] in order to achieve the thermoelectric cooling and higher temperature operation of THz QCLs. For this kind of modulation well-barrier height modulation height active structure. It is also expected to utilize the further indirect injection design.

The recent two methods are introduced for improving the performance of THz QCLs. We demonstrate a relative compact-size semiconductor THz source unit by the recent fabrication of THz QCL devices for real applications. The operation temperature largely limited the real compact size potable THz applications by semiconductor-based QCLs. A compromise solution is to capitalize on the liquid nitrogen (LN_2) cooling Dewar condenser, keep the useful characteristics of THz QCLs and reduce the cooling system size, and realize the robust portable compact size THz source unit by QCLs. The output power is one of the most important characteristics required for the different real THz applications. The recently large peak output power THz QCLs are realized by a large size mesa with a semi-insulated surface plasmon (SI-SP) waveguide and achieved the 1 W peak power. But this still limited on the pulse operation, the average output of THz QCLs still recorded by the previous report [13]. Here, we introduce a Dewar condenser cooling system with our recent fabricated metal-metal waveguide (MMW) modulation height active structure THz QCLs and the premier measurement results of few microwatts average power with milliwatt order peak power, larger output power THz QCLs with the peak output power of 250 mW and the average output power of 2.2 mW, indicating the further improvement in the direction of the continuous-wave (CW) mW order average power operation. In the following paragraphs, we introduce the details of indirect injection low frequency THz QCLs, modulation height active structure design, and a Dewar condenser type THz source unit by THz QCLs.

2. Indirect injection THz QCLs for low frequency high T operation

An indirect injection scheme was first reported for mid-IR QCLs among the different kinds of injection schemes [14]. Illustrative band diagrams of typical resonant tunneling injection schemes are shown in **Figure 1**: (a) for a simplified three-level system and (b) for a simplified four-level indirect scattering-assisted injection scheme with diagonal THz emission, which is used in this work. Improvements in performance have been successfully achieved using fast and smooth LO phonon injection without carrier accumulation. This injection scheme is also expected to circumvent the limitations of the RT injection scheme and to enable the realization of low-frequency THz QCLs operating at high temperatures. However, the much larger radiative energy in the mid-IR region causes different subband transport characteristics that need to be considered in designs for THz frequencies. It is difficult to implement an indirect injection design with the correct carrier injection and narrow radiative energy at THz frequencies. Even this design scheme still suffers difficulties in the THz region compared with mid-IR QCLs. After several theoretical proposals [15–17], a few recent experimental reports [8, 18–20] of indirect injection designs show promising results and demonstrate that there is high potential for attaining low-frequency, high-temperature THz QCLs.

Here, we demonstrate an $Al_{0.175}Ga_{0.825}As/GaAs$ QCL design that uses a combination of indirect injection and a less vertical diagonal for emission in the THz region. Structures were grown on semi-insulating GaAs (100) substrates by solid-source molecular beam epitaxy (MBE). The growth sequence started with a 250-nm-thick $Al_{0.6}Ga_{0.4}As$ etch-stop layer. The active/injection layers, which are sandwiched between two 100-nm-thick n-GaAs layers ($n = 5.0 \times 10^{18}$ cm^{-3}), are repeated 185 times with thicknesses of **3.95**/7.99/**2.16**/9.02/**3.19**/6.86/**3.76**/*14.38* (nm), and with $n = 2.1 \times 10^{16}$ cm^{-3} modulation doping employed for the widest well. After MBE growth, the wafers were processed into Cu-Cu MMW by Au-Au thermo compression wafer bonding. The 170 μm-wide ridges of the laser bars were fabricated using photolithography and Cl_2 dry etching. Finally, we cleaved the laser bars with a cavity length of 1.5 mm and measured them using a cryogenic Si-bolometer with an FT-IR system in pulsed mode. The emission spectrum is measured at 5 K and lasing is at 7.85 meV, corresponding to a frequency of 1.89 THz (wavelength of 158 μm). The observed emission energy was a little higher than the designed energy, possibly due to the grown layers being 0.5% thinner and the measured large operating voltage due to the decrease in voltage at the Schottky contacts on either side of the metal-metal waveguides. The current density-light output characteristics (*I-L* curve) and the current density-voltage characteristics (*I-V* curve) of this 1.9 THz $Al_{0.175}Ga_{0.825}As$/GaAs QCLs indicate that the *J*th at low temperature is about 700 A/cm^2 with an output peak power in the range of a few milliwatts, which was calibrated from the measurements conducted by using the Si-bolometer. In the low output range near the laser threshold, our *I-L* curve shows large noise during the measurement. This might be due to the experimental setup and the very low duty cycle (0.02%) pulse measurement. This THz QCL operated up to 160 K at a frequency of 1.9 THz, which is 1.8 times higher than the empirical temperature-frequency limitation of the conventional RT injection scheme.

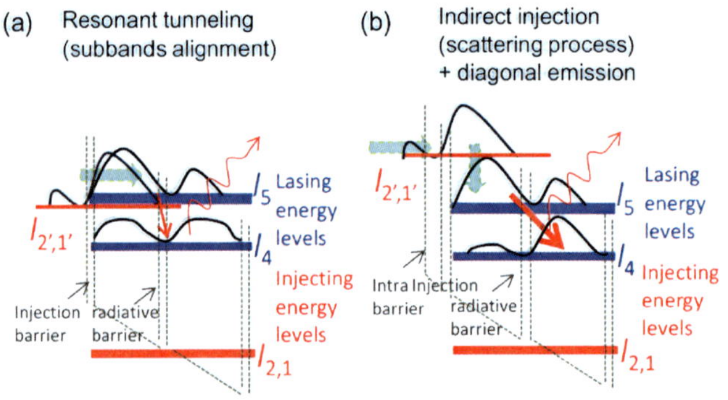

Figure 1. Illustrative band diagrams of (a) resonant tunneling injection scheme with subband alignment and (b) indirect scattering-assisted injection scheme with diagonal THz emission.

In practice, this kind of simplified four-level indirect injection scheme depends mainly on the external electrical field. **Figure 2** shows the dependence of the energy separation of the subbands on the external electric field for our structure. In the low electric field region around 12–14 kV/cm, the energy separation and band-diagram conditions are similar to the conventional RT injection scheme, as shown in **Figure 3(a)**: it shows a simplified three-level-type design including two lasing levels and one injection level with a subband alignment injection process; the band diagram related to this is shown in **Figure 1(a)**. The injection process occurs from $l_{2'}$ (the injection level at low bias) to $l_{1'}$ (the upper lasing level at low bias) via an RT injection process. The emission frequency of about 5.0 THz (~20 meV) between $l_{1'}$ and l_5 (lower lasing level at low bias) is similar to the conventional RT injection lasing conditions. The first reported indirect type scattering-assisted (SA) injection QCLs in the THz region [8] also, initially, exhibited lasing at 4 THz. This emission occurred between similar subbands, $l_{1'}$ and l_5, to those in the RT injection scheme. When the external electric field is increased, the mechanism between $l_{1'}$ and l_5 changes from a radiative process to an injection process by fast nonradiative thermally activated LO phonon scattering. At higher electric fields, an LO phonon scattering process injects carriers and assists the targeted depopulation between l_5 and l_4 for emission at 1.8 THz. Here, our design optimizes the energy separation between $l_{2',1'}$ and l_5, increasing it in order to approach the designed LO phonon energy. The fast nonradiative-injection scattering process between $l_{2',1'}$, and l_5 occurs earlier and restricts the low electric field emission to 4–5 THz. We also carefully optimized the wavefunctions between l_1 and l_5, so that there is less overlap at low electric fields, with the related small oscillator strength reducing the possibility of lasing.

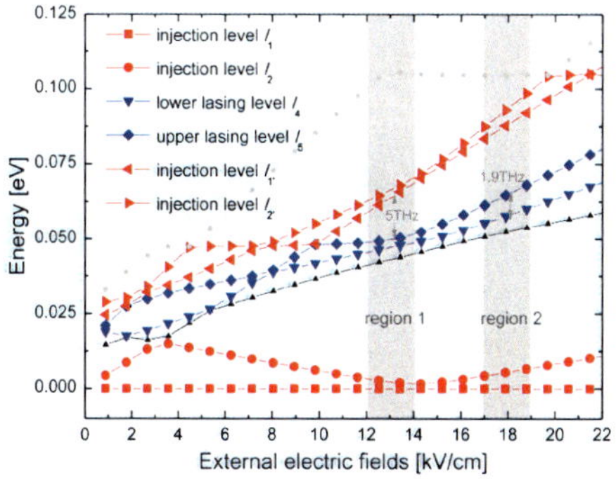

Figure 2. Calculated energies of the subband levels in the band diagram of the designed structure with respect to the external electric field. Region 1 is similar to conventional THz QCLs using the RT injection scheme, with a radiative energy of around 5 THz. Region 2 is for the indirect injection scheme combined with diagonal emission with a radiative energy below 2 THz.

We focus only on the realization of a radiative process at high electric fields for 1.9 THz lasing. This occurs in the range of 17–19 kV/cm, as shown in **Figure 2**, which is related to the designed band diagram presented in **Figure 3(b)**. Here, we utilize the LO phonon scattering process to selectively inject carriers from previous injection/extraction levels ($l_{2',\ 1'}$) to the upper lasing level (l_5) indirectly. Then the LO phonon depopulation scheme extracts the carriers by fast LO phonon scattering from the lower lasing level (l_4) to the injection level ($l_{2,1}$) and injects the carriers into the next period of the cascade in order to achieve population inversion within l_5 and l_4. We calculated the relationship between the injection selectivity and the overlap between the wavefunctions of l_5 and l_4; as the spatial separation between the wavefunctions increases, the THz radiation changes from vertical to more diagonal. The indirect injection selectivity from the injection level to the two lasing levels improves. As the overlap of the $l_{2',\ 1'}$ and l_4 wavefunctions gets smaller, the unwanted parasitic leakage channel between $l_{2',1'}$ and l_4 is reduced. As calculated, if the LO-phonon scattering relaxation time from $l_{2',1'}$ to l_5 is kept near 0.7 ps, and the wavefunction overlap ratio of $f(l_{2',1'}, l_4)/f(l_{2',1'}, l_5)$ is changed from 0.9 (large overlap of l_5 and l_4) to 0.18 (small overlap of l_5 and l_4 used in this design), then the emission of l_5 to l_4 changes from vertical to diagonal. The relaxation time for the parasitic injection scattering from $l_{2',1'}$ to l_4 increases from 0.8 to 4 ps. It estimates the improvement of injection selectivity by increasing radiation diagonal and realizes this indirect injection scheme in THz QCLs. In our previous work, we demonstrated that the use of AlGaAs with a high Al content in a THz QCL [21] brings some advantages for high temperature operation. This high-Al content barrier structure provides the benefit of having high conduction-band offsets with spatially separated wavefunctions. We prefer to use thin high-Al content AlGaAs barriers or different barrier height active structures (it will describe in detail in the next paragraph) in order to achieve an indirect injection scheme at THz frequencies with better injection selectivity by utilizing the diagonal emission design.

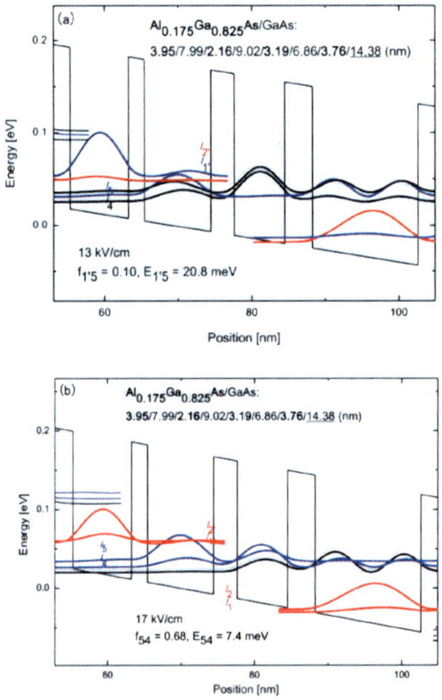

Figure 3. Self-consistent calculations of the conduction band diagrams of the $\mathbf{Al_{0.175}Ga_{0.825}As}$/GaAs THz QCLs in two different external electric fields: (a) 13 kV/cm and (b) 17 kV/cm. The blue subbands are the designed lasing levels and the red ones are the injection levels. The layer thicknesses are **3.95**/7.99/**2.16**/9.02/**3.19**/6.86/**3.76**/*14.38* (nm). The widest well is doped at a sheet density of $n = 2.95 \times 10^{10}$ cm^{-2}.

This combination of injection with the emission process also overcomes the critical injection barrier thickness in the RT injection scheme, and achieves a wider dynamic range for the operating current density. In the RT injection design, the dynamic range of the operating current density is related to the emission energy of the QCLs because the resonant condition

http://dx.doi.org/10.5772/65351

for lasing corresponds to the subband alignment from $l_{2',1'}$–l_4 to $l_{2',1'}$–l_5 [8]. The radiative energy at low frequencies (~8 meV for 2 THz) is similar to the typical energy broadening of the subbands, which is in the range of a few meV. This causes difficulties for selectively and correctly injecting carriers from $l_{2',1'}$ to l_5. A wider injection barrier can maintain superior injection selectivity and restrain the $l_{2',1'}$–l_4 leakage current, but also makes the injection transport incoherent and inefficient [22, 23], and reduces the strength of the emission oscillator. The realization of large population inversion (optical gain) and injection current are somehow mutually exclusive in design utilizing the RT injection scheme. This means the optimization of the injection barrier thickness is critical during the design of the structure [24]. We optimize the 2.16 nm thickness of injection barriers to give the lager leakage than the injection, and cannot achieve the required lasing condition at this electric field with simplified three-level band diagram. This injection barrier thickness is also considered for the lasing conditions in our simplified three-level system with higher external electric fields.

When we replace the injection process by a scattering process, as shown in **Figure 1(b)**, the injection from $l_{2',1'}$ to l_5 is not directly related to the RT process. These two levels are energetically separate. The RT current is not limited by the thickness of the injection barrier, and we have a wider optimization space for thinning the intrainjection barrier using the scheme shown in **Figure 3(b)** (the function of the intrainjection barrier is similar to that of the injection barrier in the RT injection scheme shown in **Figure 3(a)**: the current flows to the injection level $l_{2',1'}$ through this barrier by the RT process). This can maximize the resonant injection current flow across the cascade structure, achieving smooth and rapid population inversion at low frequencies. We also calculated the dependencies on the injection barrier and emission barrier thicknesses of the emission ($l_5 \geq l_4$), injection ($l_{2',1'} \geq l_5$), and leakage ($l_{2',1'} \geq l_4$) between the subbands. This data is used to optimize the optical gain (emission) and to smooth the injection current at 17 kV/cm (region 2 in **Figure 2**) using the indirect injection scheme in the simplified four-level band system. The dynamic range of the operating current density, as shown in **Figure 3(b)**, is also not directly limited by the subband alignment between $l_{2',1'}$ and l_4 to l_5. Carrier injection from $l_{2',1'}$ to l_5 by the LO phonon scattering process has greater tolerance to the electric field and thermal perturbations. It provides a wider dynamic range for the current density even at narrow radiative energy.

Here, we demonstrate 1.9 THz $Al_{0.175}Ga_{0.825}As$/GaAs QCLs with T_{max} up to 160 K, operating at temperatures much higher than the traditional empirical $h\omega/k_BT$ limitation for RT injection at low frequencies (T_{max} ~1.8 $h\omega/k_B$). The structure utilizes an indirect scattering process for injection and a high Al-content barrier in order to achieve a larger spatial separation between the wavefunctions and more diagonal THz emission. It is also easier to realize an indirect injection scheme in the THz region using the AlGaAs/GaAs material. Moreover, this design has greater tolerance to both the electric field and thermal perturbations with a smooth current flow. We estimated that this kind of combination design scheme has better selectivity relative to the correct level of injection. With careful optimization of the wavefunctions, we realized lasing at frequencies below 2 THz only. With the structure optimized in different directions, this design scheme for THz QCLs is a promising means for extending QCLs to lower frequencies and also to higher temperature operation. The further design of these indirect injection

THz QCLs is possible by the larger design freedom modulation height barrier-well active structure design, which we will describe in the next paragraph.

3. Modulation barrier-well active structure design THz QCLs

In order to design the further indirect injection THz QCLs with more design freedom in narrow energy THz frequency, which traditionally usual suffered from opposite thickness optimization. Here, we introduce the active structure design with different heights of barriers and wells. First, we present primary experimental results by arranging the recently recorded three-well resonant tunneling structure design; modulate the Al the height of barriers with the different emission barrier compositions and introduce an external thin barrier in the widest extraction/injection well. This modulation barrier THz QCL sample succeeds lasing at 3.7 THz with the maximum operation temperature of up to 145K.

Currently, the best performing designs are mainly based on the resonant phonon depopulation scheme for extraction utilizing the AlGaAs/GaAs material systems with one kind of Al barrier concentration 15%. The modulation Al composition barrier THz QCL samples are grown by solid-state MBE with multiple Al cells. The thickness and Al compositions are precisely controlled within 1% difference. Then the devices are fabricated in the Cu-Cu metal-metal waveguide by photolithograph and dry etching. The active region structure design and band diagram are shown in **Figure 4**.

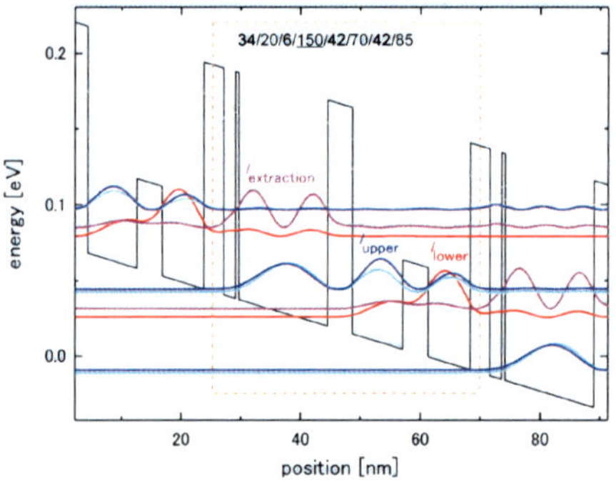

Figure 4. Band diagram of our modulation barrier THz QCLs in this paper. Based on the three-well resonant tunneling design and arranged with one external thin barrier and lower emission barrier.

First, we follow the three-well resonant tunneling injection design [7], achieve diagonal emission and increase the spatial separation between upper and lower lasing levels in order to reduce the parasitic injection leakage comes from nonradiative thermally activated longitudinal phonon scattering at high temperature operation. At the widest extraction/injection well, we expected to add an external thin and high layer in order to improve extraction by interface roughness (IR) scattering, which is one solution used in mid-infrared QCL design

http://dx.doi.org/10.5772/65351

[25]. The simulated Al composition of this external barrier with an amount of IR scattering, oscillator strength between the extraction/injection level (l_{ext}) to upper lasing level (l_{upper}) and the lower lasing level (l_{lower}) is shown in **Figure 5**. The extraction from the l_{lower} to l_{ext} can induce by the fast elastic IR scattering comes from this thin external barrier with the increase of barrier height; the extraction/injection efficiency can be improved by increasing the barrier height, as shown at the circle point in **Figure 5**. The related oscillator strength of l_{ext} to l_{upper} and l_{lower} (square and triangle points) is decrease and increase with the increase of the barrier height inversely. It means that the wavefuctions of l_{ext} are more localized in the extraction/injection well; the plastic leakage from the l_{upper} to l_{ext} can be reduced. In the same time, maintain the strong extraction from l_{lower}. Finally, the injection selectively and efficiency can be improved in the resonant tunneling scattering design. But the too thin layer causes the difficulty to the growth thickness control.

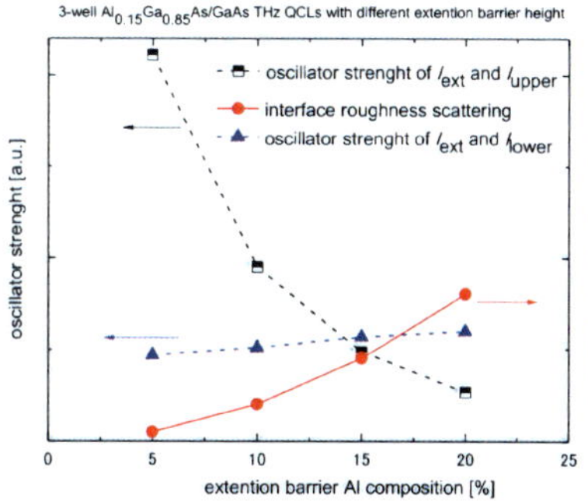

Figure 5. Simulation of external barrier Al composition with the amount of IR scattering, oscillator strength between the extraction/injection level (l_{ext}) to upper lasing level (l_{upper}) and the lower lasing level (l_{lower}).

For the further design, we also would like to increase all the barrier compositions, so here we first keep this thin barrier composition the same with others barriers at 15%. The lower emission barrier between two emission wells is modulated in order to reduce the nonradiative IR scattering at the emission layers. The Al composition of the emission barrier with the amount of IR scattering and oscillator strength between two lasing levels (l_{upper}, l_{lower}) is calculated as shown in **Figure 6**. The amount of IR scattering is proportional to the square of barrier height (Al composition 7.5% barrier in **Figure 4**). Reduction of barrier height is an efficient way to exclude this nonradiative scattering in the emission region. The utilization of a lower emission barrier height of 7.5% is to keep the oscillator strength same as the original design and reduces the IR scattering. The design $\mathbf{34}_{(0.15)}/20/\mathbf{6}_{(0.15)}/150/\mathbf{42}_{(0.15)}/70/\mathbf{42}_{(0.075)}/85$ $\mathbf{Al_xGa_{1-x}As_{(y)}}$/GaAs modulation THz QCLs are measured by the cryogenic cryostat with the Si-bolometer and FT-IR system in pulse mode. The maximum operation temperature (T_{max}) is up to 112 K with a threshold current density of 0.85 kA/cm^2 and wide operation current density dynamics range about 1 kA/cm^2. The 3.7 THz peak is operated up to the T_{max} of this device. This frequency is consistent with the design energy separation of 15.5 meV between two lasing levels under designed operation electric fields of 12.6 kV/cm. But at low operation temperature below 90

K, one more unexpected peak also estimates at around 3.4 THz. This multiple peaks can be explained by the transition from both $l_{upper(1,2)}$, as shown in **Figure 4**. The similar wavefunctions and oscillator strength to l_{lower} ($f_{upper1\text{-->}lower}=0.26$ and $f_{upper2\text{-->}lower}=0.23$) with an energy separation of about 1.2 meV (0.3 THz) of two upper lasing levels during the designed electric field. This energy separation is consistent with the frequency difference between two peaks. The wider operation current density dynamics range also can be explained by the combination of two lasing *I-L* characteristics of maniple lasing. These might come from the structure not well optimized near the emission levels and reduced the total injection efficiency to the designed upper lasing level and resulting in the lower temperature performance as we expected for this primary modulation barrier THz QCLs.

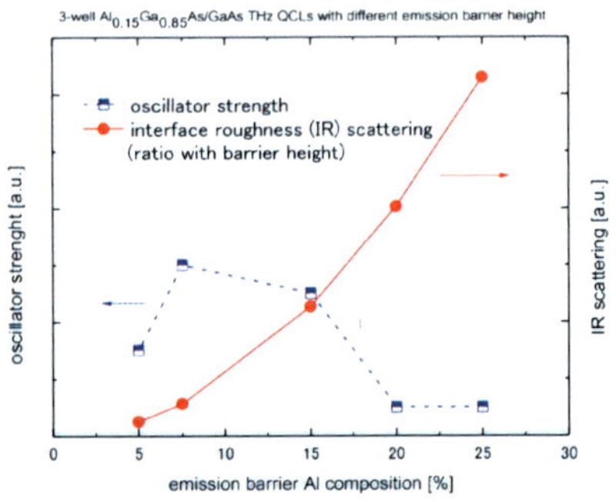

Figure 6. Simulation of emission barrier Al composition with the amount of IR scattering, oscillator strength between two lasing levels (l_{upper}, l_{lower}).

We utilize the lower emission barrier to reduce the IR scattering at the emission region, add an external thin barrier to improve the injection efficiency and reduce plastic leakage current from l_{upper} to l_{ext}. We demonstrate the lasing of this kind of modulation barrier design structure but still show the poor temperature performance lower than we expected. The active structure from this approach, as the result, is similar to the other theoretical design of THz QCLs with different barrier height [12]. On the basis of initially experimental results, we further optimize the thickness of emission barriers in order to deaden the multiple lasing and improve the temperature operation at the designed 3.7 THz. Then obtained the expected 3.7 THz operation only with the same threshold current density, and finally give T_{max} to 145 K. The future structures optimization and utilize the indirect injection scheme combine with high Al composition design in injection layers with low Al composition in emission layers. We expect the realization of thermoelectric cooling temperature operation THz QCLs.

4. Relatively compact-size potable condenser type THz source by QCLs

After introduce the recent fabricated THz QCLs toward high temperature operation at low frequency (<2 THz) and the 3–4 THz region. For THz applications, the output power is more

http://dx.doi.org/10.5772/65351

critical than the operation temperature. Even the recent devices cannot operate at the room temperature. Utilizing some kinds of cryogenic systems, it is still possible to achieve the relatively compact size potable THz application by THz QCLs. Here, we introduce the liquid nitrogen Dewar condenser to combine the fabricated THz QCLs as mentioned above. The size of Dewar is 14 cm^2 with 28 cm high without the outside external power supply (**Figure 7(a)**), the QCL sample array (**Figure 7(c)**) is fixed under vacuum conditions with cooling holder direct cooled by heat conduction from LN_2. The QCL samples are combined with hyperhemispherical Si lens in front of operated mesa and an adjustable inside parabolic mirror in order to focus the output of THz QCLs (**Figure 7(d)**) and give the near collimated THz wave outside the measurement Tsurupica windows from the widely diverged MMW QCLs.

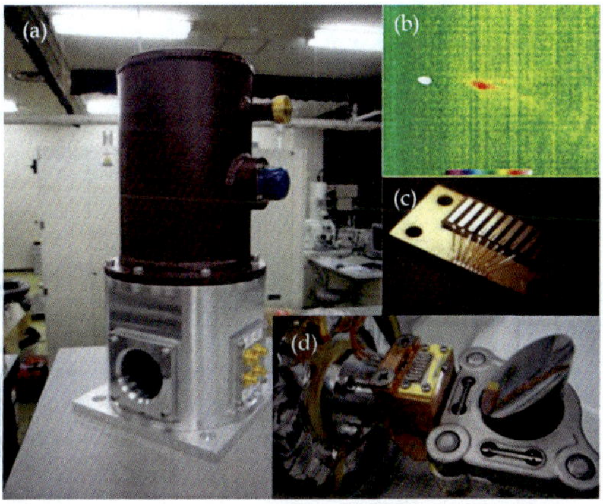

Figure 7. (a) The photograph of the Dewar condenser. (b) Far-field pattern of the Dewar condenser measured using NEC THz camera IR/V-T0831. (c) The photograph of fabricated THz QCL array with 9 mesas. (d) The fixed QCL array with hyperhemispherical Si lens and an adjustable parabolic mirror inside.

In our case, without the hyperhemispherical Si lens, the far-field pattern angle of MMW QCLs is about 40°. If coupled with the Si lens, it is possible to adjust the maximum output near 0°. This is very helpful for the adjusting the THz wave from the small size measurement window and use for real applications. The far-field pattern is also shown in **Figure 7(b)**, which measured using the NEC THz camera IR/V-T0831. The adjustment of the position of Si lens is also quite important for the increasing the output power, with and without the Si lens, the measured out power can be increased about 2.5–3 times. The active region design is based on the four well-resonant tunneling injection LO depopulation scheme with the variable barrier height modulation active structure THz QCLs as discussed above, which originally toward the high temperature operation. The waveguides also use the high temperature operation MMW. The output peak power is 0.1 mW with a duty cycle of 0.02% (1 kHz repeated frequency with 200 ns pulse width). When the duty cycle increases to 5% (10 kHz repeated frequency with 5 μs pulse width), the device gives the 37 μW average power with >1 mW peak power. The changes in duty cycle by changing the repeated input pulse frequency and pulse width are related to the peak output power of THz QCLs when the pulse width is below 1 μs. The too short pulse did not supply the enough input energy for devices. The peak power is become stable when the input pulse is larger than 1 μs. But the average power can be increased continuously with the increasing duty cycle.

Recently, our stable pulse generator (AVTECH AVO-6HZ-B) with impedance match can only operate up to 5%. The device can possibly operate at higher duty cycle with larger average output power. But when the duty cycle increases, the heat generated from the device is quite large during the operation, comparing with the measurement condition 0.02%, even with the use of continuous flow liquid helium cryostat. The device heat sink temperature is increased up to 30 K during the operation. For the Dewar case, the duty cycle is larger than 1%. The recent contact electrode by indium ball and 0.07 mm-thin cupper wire for electric connection is easy to be melted. It causes the difficulty for stable CW operation. The measurement spectrum and the stable operation of current density-voltage, current density-light output characteristics direct from the Dewar condenser, are shown in the inset of **Figure 8**. We can find that the Dewar gives the clear single color 3.8 THz lasing spectrum, stable milliwatt order peak power and microwatts order average power. The maximum peak power is 3.1 mW and maximum average power is 6.2 μW under recent Dewar setting conditions.

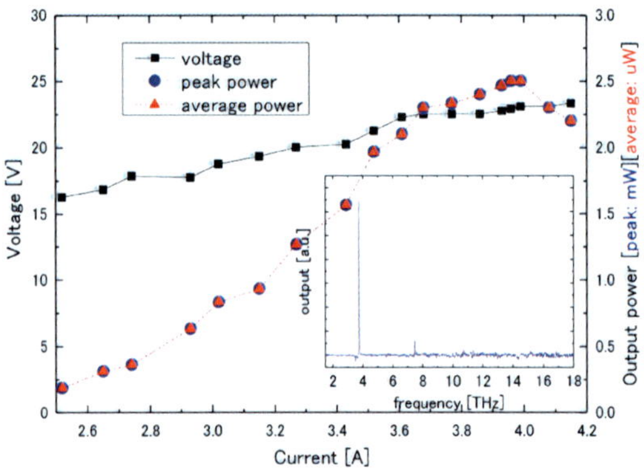

Figure 8. The measurement lasing spectrum and the stable operation of current density-voltage, current density-light output characteristics direct from the Dewar condenser.

For the CW operation with milliwatt average output power, the improvement for the recent Dewar system mainly obtained by two directions: first is the THz QCL device, recently the QCLs used in this condenser are toward the high temperature operation MMW QCLs with narrow ridge width and small mesa size. The output performance of THz QCLs generally decreases when the operation temperature is above 100 K. Upon cooling with the LN_2, the performance do not dramatically change to a large extent under low temperature conditions. The few tens K temperature operation is not important for this Dewar setting. The device fabrication move to the larger mesa size SI-SP waveguide can improve the output performance with the better far-field pattern. It is suitable for the high output 77 K Dewar condenser with

http://dx.doi.org/10.5772/65351

the easier optical aliment in applications. Second direction is the improvement and consideration of cooling efficiency of the Dewar condenser. The recent operation duty cycle is limited by the wire connection and the electrode of the mounted sample holder not by the device itself. If we can reduce the temporary heat increase with large duty cycle by wide size heat conduction and good thermal conductivity, it is possible to improve the operation duty cycle of the Dewar condenser.

On the basis of the first direction, we introduce the recent fabricated modulation active structure THz QCL cooling and setting with the active structure design by more vertical emission and large current injection. This way we achieve the large output THz QCLs with a peak power of 250 mW and an average power of 2.2 mW under the condition of duty cycle 1%. L-I-V characteristics, lasing spectrum, and far-field pattern measured using THz camera are shown in **Figure 9**. The LN_2 Dewar condense demonstrated as a robust compact size THz source unit with stable milliwatt order peak power and microwatts order average power operation by THz QCLs. And the large output power devices without high temperature operation (>100 K) also show the potential for the more powerful THz source unit toward the target of CW operation with milliwatt average output power. For further improvement of Dewar condenser, we replace the recent fabrication by SI-SP waveguide with large mesa size and high output power active region design, increase the current duty cycle limitation comes from the power supply, cooling efficiency of connect wire and electrode.

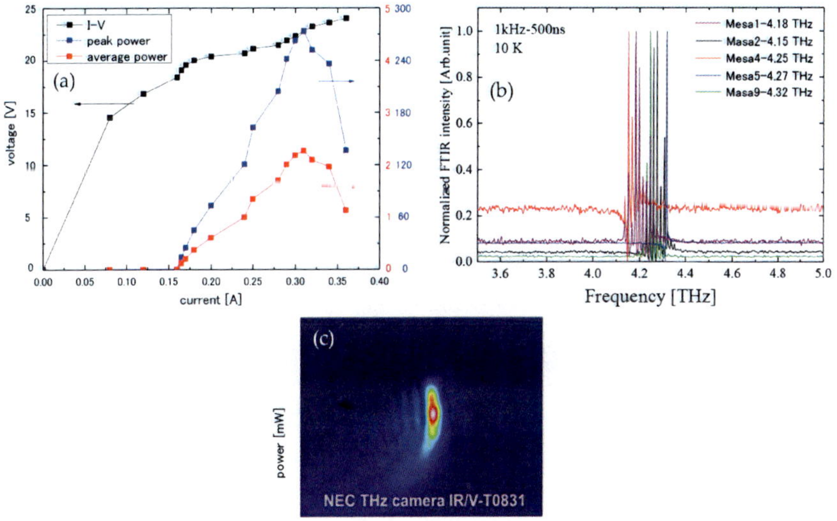

Figure 9. The measurement of (a) current density-voltage, current density-light output characteristics, (b) lasing spectrum, and (c) far field pattern of large output THz QCLs.

Here, we discuss the design and fabrication of THz QCLs toward the high temperature and large average output power operations for the real THz applications with the relatively compact portable size cryogenic cooling systems. We also describe the temperature performance parameters of THz QCLs, introduce the results of an indirect injection design scheme in the THz region and modulation height active structure design with different barriers and wells for further design direction. The recent fabricated THz QCLs are combined with the liquid nitrogen cooling Dewar condenser to demonstrate the relatively compact potable THz source unit by QCLs. The different injection schemes in THz and barrier-well height design in the active region introduce one of the directions for the further high temperature and large output power operation of THz QCLs for real applications.

Author details

Tsung-Tse Lin

Address all correspondence to: ttlin@riken.jp

RIKEN Center for Advanced Photonics, Terahertz Quantum Device Research Team, Sendai, Japan

References

[1] T. K.-Ostmann and T. Nagatsuma: J. Infrared. Milliw. TE. 32 (2011) 143.

[2] J. Faist, F. Capasso, D. L. Sivco, C. Sirtori, A. L. Hutchinson, and A. Y. Cho: Science 264 (1994) 553.

[3] R. Köhler, A. Tredicucci, F. Beltram, H. E. Beere, E. H. Linfield, A. G. Davies, D. A. Ritchie, R. C. Iotti, and F. Capasso: Nature 417 (2002) 156.

[4] C. Walther, M. Fischer, G. Scalari, R. Terazzi, N. Hoyler, and J. Faist: Appl. Phys. Lett. 91 (2007) 131122.

[5] C. W. I. Chan, Q. Hu, and J. L. Reno: Appl. Phys. Lett. 101 (2012) 151108.

[6] L. Li, L. Chen, J. Zhu, J. Freeman, P. Dean, A. Valavanis, A.G. Davies, and E. H. Linfield: Electron. Lett. 50 (2014) 309.

[7] S. Fathololoumi, E. Dupont, C. W. I. Chan, Z.R. Wasilewski, S.R. Laframboise, D. Ban, A. M'aty'as, C. Jirauschek, Q. Hu, and H. C. Liu, Opt. Express 20 (2012) 3866.

[8] S. Kumar, C. W. I. Chan, Q. Hu, and J. L. Reno: Nat. Phys. 7 (2011) 166.

[9] T.-T. Lin, L. Ying, and H. Hirayama: Appl. Phys. Express. 5 (2012) 012101.

[10] C. W. I. Chan, Q. Hu, and J. L. Reno: Appl. Phys. Lett. 103 (2013) 151117.

[11] G. Scalari, M. I. Amanti, M. Fischer, R. Terazzi, C. Walther, M. Beck, and J. Faist: Appl. Phys. Lett. 94 (2009) 041114.

[12] A. Matyas, R. Chashmahcharagh, I. Kovacs, P. Lugli, K. Vijayraghavan, M. A. Belkin, and C. Jirauschek: J. Appl. Phys. 111 (2012) 103106.

[13] B. S. Williams, S. Kumar, Q. Hu and J. L. Reno: Electron. Lett. 42 (2006) 89.

[14] M. Yamanishi, K. Fujita, T. Edamura, and H. Kan: Opt. Express. 16 (2008) 20748.

[15] H. Yasuda, T. Kubis, P. Vogl, N. Sekine, I. Hosako, and K. Hirakawa: Appl. Phys. Lett. 94 (2009) 151109.

[16] T. Kubis, S. R. Mehrotra, and G. Klimeck: Appl. Phys. Lett. 97 (2010) 261106.

[17] T. Liu, T. Kubis, Q. J. Wang, and G. Klimeck: Appl. Phys. Lett. 100 (2012) 122110.

[18] E. Dupont, S. Fathololoumi, Z. R. Wasilewski, G. Aers, S. R. Laframboise, M. Lindskog, S. G. Razavipour, A. Wacker, D. Ban, and H. C. Liu: J. Appl. Phys. 111 (2012) 073111.

[19] K. Fujita, M. Yamanishi, S. Furuta, K. Tanaka, T. Edamura, T. Kubis, and G. Klimeck: Opt. Express 20 (2012) 20647.

[20] S. G. Razavipour, E. Dupont, S. Fathololoumi, C. W. I. Chan, M. Lindskog, Z. R. Wasilewski, G. Aers, S. R. Laframboise, A. Wacker, Q. Hu, D. Ban, and H. C. Liu: J. Appl. Phys. 113 (2013) 203107.

[21] T.-T. Lin and H. Hirayama: Phys. Stat. Sol. C 10 (2013) 1430.

[22] C. Sirtori, F. Capasso, J. Faist, A.L. Hutchinson, D. L. Sivco, and A. Y. Cho: IEEE J. Quantum Electron 34 (1998) 1722.

[23] S. Kumar and Q.Hu: Phys. Rev. B 80 (2009) 245316.

[24] H. Callebaut and Q. Hu: J. Appl. Phys. 98 (2005) 104505.

[25] Y. Chiu, Y. Dikmelik, J. B. Kurghin, and C. Gmachl: in the 11th International Conference on Intersubband Transitions in Quantum Wells, Sardinia, Italy (2011).

Chapter 5

Power Amplification and Coherent Combination Techniques for Terahertz Quantum Cascade Lasers

Yan Xie, Yanfang Li, Jian Wang, Ning Yang, Weidong Chu and Suqing Duan

Additional information is available at the end of the chapter

http://dx.doi.org/10.5772/65350

Abstract

Power amplification and coherent combination are important ways to improve the output power and beam quality of single-mode terahertz quantum cascade lasers (THz QCLs). Up to date, the tapered waveguide is the most convenient way to amplify the power of THz QCLs. The self-focusing effect in tapered THz QCLs induces non-monotonic behaviours of the peak power and far-field beam divergence, which lead to the existence of optimal structural parameters. The surface and lateral grating techniques have also been employed in tapered THz QCLs to further improve the spectral purity. For coherent combinations, the progress of facet-emitting phase-locked arrays of THz QCLs is still limited due to both the lack of the understanding of dynamics of coupled QCLs and the difficulties in designing high-performance coupled wave-guides. We will briefly review the developments of coherent arrays of THz QCLs and present a design of monolithic QCL arrays with common coupled cavity to achieve the optical mutual injection, which may provide a new way for coherent combination of THz QCLs.

Keywords: terahertz, quantum cascade lasers, tapered waveguide, phase-locked array

1. Introduction

The performance achieved by state-of-the-art terahertz quantum cascade lasers (THz QCLs) [1–3] has demonstrated impressively that these compact semiconductor laser sources have become the ideal coherent radiation source for a large variety of applications, such as security check, free space optical communication, terahertz spectroscopy and imaging, particularly when considering the tremendous degree of customization feasible due to the comprehensive design freedom for both the gain material and the device geometry [4]. To date the peak output power of THz QCLs can exceed 1 W. However, the high-power devices are usually multi-mode operated that strongly limits their applications where the purity of spectrum is desired. The power amplification and coherent combination techniques may provide ways to improve the output power of THz QCLs and meanwhile maintain the single-mode operation. Besides, the beam divergence of a THz QCL is quite larger and the beam quality is also worse than that of mid-infrared QCLs or traditional semiconductor lasers since its aperture is comparable with the wavelength. Some of the power amplification techniques and the coherent combination techniques may also help to improve the brightness and the beam quality of THz QCLs.

As a method of power amplification, the tapered gain region can effectively improve the output power of a laser maintaining the single-mode operation, which has been widely employed in diode lasers and mid-infrared QCLs. For THz QCLs, the tapered cavity also increases the aperture, which will decrease the divergence of the laser beam in the horizontal direction.

Besides the power amplification, the coherent power combination of lasers is also a useful way to improve the output power maintaining the purity of the lasing spectrum. The phase-locked arrays of surface-emitting THz QCLs have demonstrated the single-mode operation as well as the great reduction of the divergence of the laser beam. However, due to the difficulty of design and fabricate the high efficiency and low-loss coupling waveguide, the progress of facet-emitting phase-locked arrays is still limited.

This chapter will describe the tapered THz QCLs as a typical power amplification technique and the development of phase-locked THz QCL arrays as coherent power combination techniques. For the tapered THz QCLs, we will present the basic characteristics of tapered THz QCL devices at first and introduce the development of THz tapered distributed feedback (DFB) lasers with surface grating and lateral gratings, which further ensure the single-mode (longitude mode) operation. For the phase-locked THz QCL arrays, we will briefly review the developments of non-coherent arrays and phase-locked arrays of mid-infrared QCLs, and then study the basic dynamics of coupled THz QCLs to discuss the difficulties of developing coherent arrays for THz QCLs. Finally, we will introduce the recent developments of phase-locked THz QCL arrays.

2. Power amplification technique of tapered THz QCLs

2.1. Basic properties of a tapered THz QCL

The tapered resonant cavity has been widely adopted in near-infrared diodes [5–8] and mid-infrared QCLs [9–13]. The introduction of the tapered region may affect the propagation of the fundamental mode of the light field, threshold current and output power of laser devices.

Li et al. studied a prototype tapered THz QCL for the first time [14] and demonstrated the effects of power amplification and improvement of the beam divergence. The active region of their THz QCLs is bound-to-continuum transition design and the tapered devices were

http://dx.doi.org/10.5772/65350

fabricated with standard optical lithography, etching and bounding processes. The details of the design and the fabrication can be found in reference [14]. The SEM image of several tapered THz QCLs before cleaving and bounding is shown in **Figure 1(a)**. The devices are composed of same ridged region with 0.5 mm length and 103 μm width and different tapered angles. The narrow ridge and some elaborate design of the waveguide ensure the single-transverse-mode operation. However, the devices are still multi-longitude-mode operated especially under high driving currents, as shown in the inset of **Figure 1(b)**. The measurements of the output power have demonstrated the effect of power amplification of the tapered active region (see **Figure 1(b)**). The peak output power of tapered devices increases with tapered angles changing from 0° to 5°. However, with further increasing the tapered angle, the peak output power decreases.

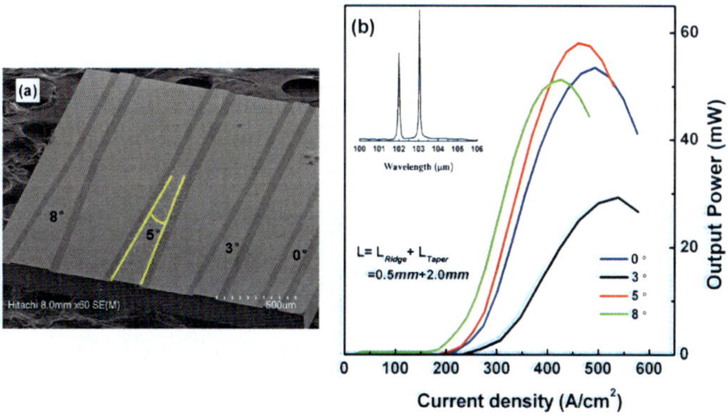

Figure 1. (a) SEM image of the tapered THz QCL with tapered angles equal to 0°, 3°, 5° and 8°. (b) Output power versus current density at 10 K of the tapered THz QCL with a total length of 2.5 mm and tapered angles 0°, 3°, 5° and 8°. The inset shows the emission spectrum of the 5°-tapered device. Citing from Ref. [14].

Another important effect of tapered THz QCLs is the reduction of the horizontal beam divergence. **Figure 2(a)** and **(b)** show the full width at half maximum (FWHM) of the far-field laser beam of the tapered THz QCL with the varying tapered angles and tapered length. The measured FWHM angles decrease from 31.72° to 19.05° as the tapered angle increases from 0° to 5° in **Figure 2(a)** and then increase while the tapered angle reaches 8°. Similarly, a non-monotonic change is also exhibited between FWHM angles and tapered region length as shown in **Figure 2(b)**.

Since the devices are carefully designed to avoid the multi-transverse-mode operation, the non-monotonic behaviours of the far field divergence of THz QCLs cannot be attributed to the emergence of high-order transverse modes. Then the abnormal increase of the divergence of the laser beam of the devices with large tapered angle or large length may be due to the self-focusing effect in high-power THz QCLs. The self-focusing effect of laser beam can be understood by considering the diffraction of a laser beam in material exhibiting an intensity-dependent (Kerr lensing effect) or temperature-dependent (thermal lensing effect) refraction index [15, 16]. The coupled-quantum-well nature of the THz QCLs can give a much higher nonlinear refractive index coefficient than the bulk GaAs. With the increase of the current density, i.e. the output power, the thermal accumulation can also greatly change the refraction index of the laser material and finally influence the device performance. The non-monotonic behaviours of the beam divergence of tapered THz QCLs can be understood by the simulations of the mode propagation within the waveguide and the far-field pattern of the optical field with the consideration of the intensity-dependent or temperature-dependent refraction index. The detailed simulation process can be found in reference [17]. In **Figure 2**, the simulated FWHM of tapered THz QCLs with different tapered angles and lengths are also presented, which agree with the experimental data accurately. It should be noted that the simulation suggests that the Kerr lensing effect may be the main reason for the non-monotonic behaviours of the beam divergence. The thermal lensing effect hardly influences the beam divergence but may greatly influence the output power of the devices.

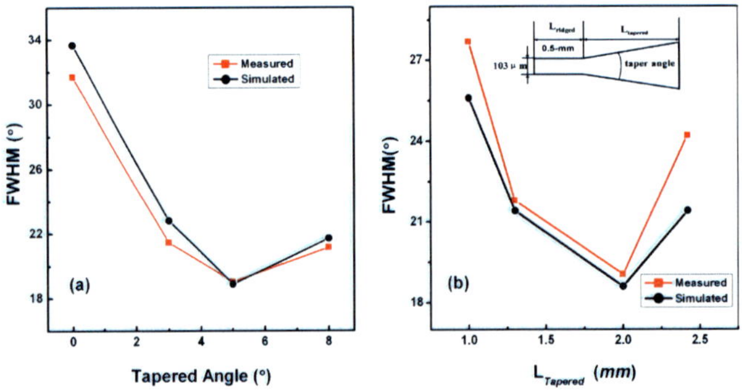

Figure 2. Measured and simulated full width at half maximum (FWHM) angles for tapered THz QCLs versus varying (a) tapered angles from 0° to 8° and (b) tapered region lengths from 1.0 to 2.42 mm. The red and black lines correspond to the measured and simulated data, respectively.

Briefly speaking, the tapered active region has prominent effects on the output powers and far-field beam of THz QCLs. Since the self-focusing effect may worsen the device's performance, there will be an optimal tapered angle for the output power and the far-field divergence. These results should be carefully considered in future imaging systems based on the tapered THz QCLs.

2.2. Single-mode techniques for tapered THz QCLs

As mentioned above, the first tapered THz QCL is only single-transverse-mode operated. To achieve the single-longitude-mode operation, further grating techniques are still required. Stable single-mode emission at a precisely designed frequency has been achieved in conventional interband lasers and in mid-infrared QCLs by using the DFB resonators. At terahertz wavelengths, the DFB QCLs are well-defined single-mode emission by means of surface grating and lateral waveguide grating. Wang et al. studied the tapered DFB THz QCLs based on metal-stripe surface grating with different grating periods of 13, 13.2 and 13.4 μm [18] (see **Figure 3**). The collected spectral peaks of 95 and 96.2 μm in **Figure 4** indicate that the single mode emission has been demonstrated for lasers with grating period at 13.2 and 13.4 μm, respectively. In this work, a high emitting optical power of 57 mW at 10 K together with a well-shaped far-field pattern is obtained by the 2.5 mm long tapered THz QCLs. Reliable single-mode emission at all injection currents and operating temperatures is realized with a side-mode suppression ratio (SMSR) > 17 dB.

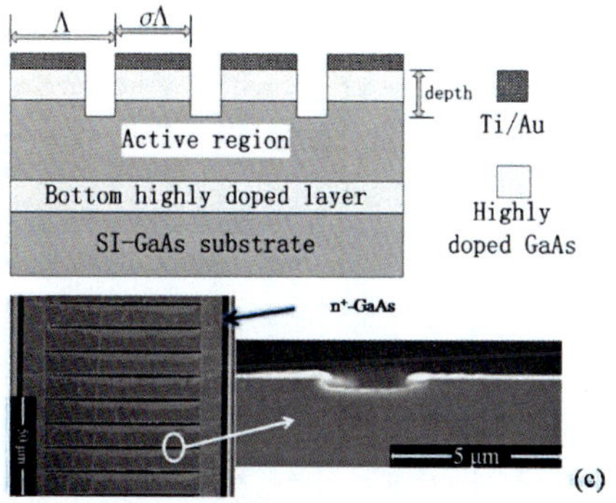

Figure 3. (a) Cross-sectional schematic view of the metal-stripe surface grating structure on a tapered THz QCL. (b) Top view of SEM image of a 13.4-μm grating. (c) Cross sectional SEM image of the selected section in (b). Citing from Ref. [18].

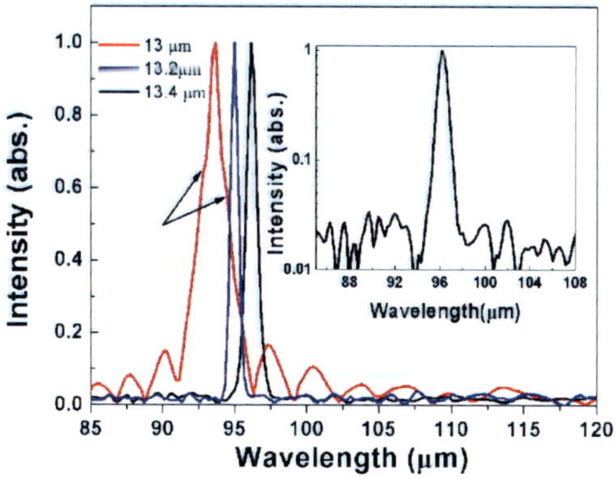

Figure 4. Emission spectra of three 2-mm-long lasers with DFB periods 13, 13.2 and 13.4 μm, respectively, were recorded in linear scale. Data were collected in pulsed mode at a 1% duty cycle with 2-μs pulses by an FTIR spectrometer with 0.5 cm^{-1} resolution at 10 K. Inset: logarithmic scale of the spectrum for the laser with 13.4-μm grating period. Citing from Ref. [18].

Using the same surface grating technique, Wang et al. also made a packaged array [19] with three different grating periods of 12.9, 13.1 and 13.3 μm to obtain a compact multi-wavelength laser source. The emission wavelengths are evenly spaced with peaks at 92.6, 93.9 and 95.1 μm, respectively. The SMSR for the longitude modes is 18 dB. The maximum peak output powers of 42, 73 and 37 mW at 10 K are realized for each individual laser. Two of lasers show pure single-transverse-mode operation with FWHM angles 21.6° and 20.3°. The rest laser (12.9 μm) exhibits minor contributions of higher-order transverse mode, but it still maintains an excellent far-field quality with a little broader angular intensity distribution.

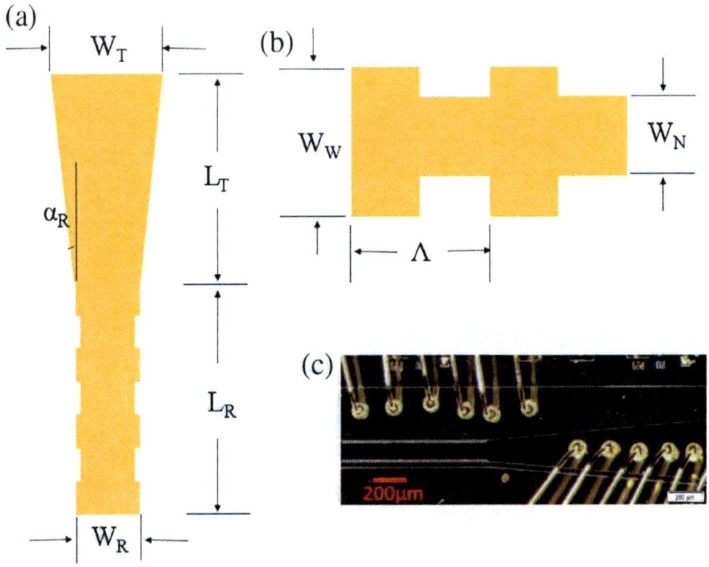

Figure 5. (a) Schematic diagram of a tapered THz QCL with lateral gratings. (b) Illustration of lateral gratings. (c) The top view of the fabricated tapered THz QCL with lateral gratings. Citing from Ref. [20].

Yao et al. obtained a single-mode tapered THz QCL with lateral grating technique [20], as shown in **Figure 5**. The designed active region is based on a bound-to-continuum transition with centre frequency of 3.4 THz. The grating period of the device is chosen to be 12.28 μm with duty cycle of 50%. The width of the ridges has been carefully designed to ensure the single transverse mode operation. As shown in reference [20], the measured maximum peak power of ~30 mW at 10 K is obtained for 2.5 mm long tapered QCLs in pulsed mode. The measured threshold current density at 10 K is to be 550 mA/cm^2 and the slope efficiency is almost 65 mW/A. As shown in **Figure 6(b)**, the emission spectrum of the tapered QCLs without lateral grating exhibits the multi-mode nature. The single-mode operation is observed for the devices with lateral grating with SMSR ~20 dB within the whole range of driving current from 3 to 3.6 A. Meanwhile, the horizontal divergence angle of the tapered THz QCLs with lateral grating is 15.5°, which is half of the devices without tapered region (32°).

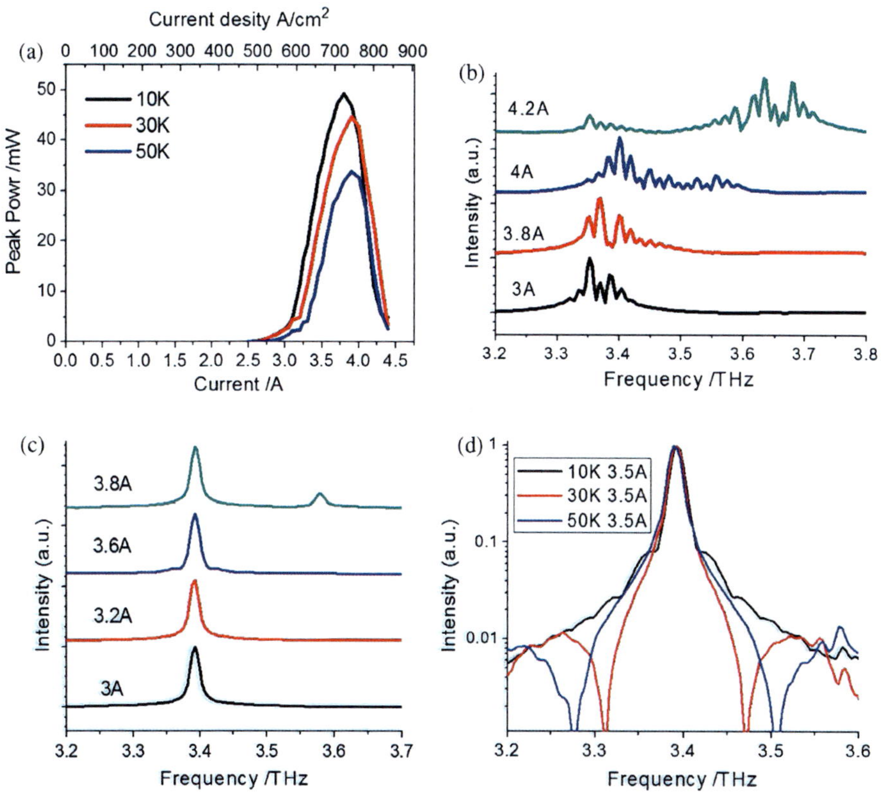

Figure 6. (a) Typical current-voltage-peak output power characteristics of 2.5 mm-long tapered QCLs. (b) Normalized emission spectra of tapered THz QCLs without lateral gratings at various injection currents in pulsed mode at the heat-sink temperature of 10 K. Spectra under four injection currents are selected: 3 A (low injection current, near the threshold current), 3.8 A (near the rollover current, for output power roll-off), 4 A and 4.2 A (the injection current far beyond the rollover current). (c) Normalized emission spectrum of tapered THz QCLs with lateral gratings at drive various injection current at the heat-sink temperature of 10 K. (d) The logarithmic plot of the emission spectrum at different temperatures with a 3.5 A driving current at which the laser achieves the maximum output power. Citing from Ref. [20].

3. Coherent power combination of THz QCLs

3.1. Brief review of QCL arrays

With the help of the monolithic technology, QCLs and their related electrical and optical components can be integrated on a chip to form QCL arrays to meet requirements of specific applications. For many real applications, such as remote sensing of chemicals, free space communication and infrared countermeasure, high-output power QCL arrays with tunable frequency range, single-mode operation and low beam divergence are highly desired.

For non-coherent arrays, the most straightforward concept is to integrate a number of single-mode QCLs, which are lasing at different wavelengths spanning the desired region with wavelength spacing smaller than the feasible range of temperature tuning. Challenges arise from reliably achieving single-mode operation at deterministic wavelength for each laser element in combination with a uniform distribution of high output power across the array and a centre-lobed far-field emission pattern. In mid-infrared region, distributed feedback QCL arrays have been capable of single-mode output continuously tunable over the whole spectral range between 8.73 and 9.43 μm, by employing a combination of electrical switching and temperature tuning [21]. However, the operation was limited by variation in output power between the individual QCLs, which was attributed to the influence of device facet on mode selection. Although the ring-cavity configuration without facets can eliminate their influence on mode selection and output power, the DFB ring-cavity surface emitting QCL arrays also face a series of disadvantages, including large-size induced beam combing challenge, minimum-centre intensity distribution [22] and low output power [23]. Besides, master-oscillator power-amplifiers [24] and wavelength beam combining approaches [25] have also been used to obtain broad spectral tenability and low beam divergence based on multi-wavelength QCL arrays. In terahertz region, continuous tunability from 1.9 to 3.9 THz at room temperature has been demonstrated in a QCL array with intra-cavity terahertz difference-frequency generation [26]. Similarly, the low and inhomogeneous output power was also a limitation on further application.

The aim of constructing coherent, i.e. phase-locked, THz array is to obtain pure super-mode emission with high beam quality. Obtaining the coherence within an array relies on the ability to control the coupling between individual lasers. Inspired by concepts developed in more mature, shorter wavelength diode laser systems, several coupling schemes, including coupling through exponentially decaying fields outside the high index dielectric core (evanescent-wave coupled) [27–29], through feedback from external reflectors (diffraction-wave coupled) [30], connecting two ridges to one single-mode waveguide (*Y* coupled or tree coupled) [31–35], through lateral propagating waves (leaky-wave coupled) [36–38] and combining graded-photonic-heterostructure (GPH) QCLs with a ring resonator [39], have exhibited excellent performance on phase-locking QCL arrays, especially in mid-infrared range.

In 2011, Faist and his co-workers presented for the first time the realization of buried heterostructure phased-locked arrays on mid-infrared QCLs. The array elements have been designed in order to maximize the overlap of the mode with the active medium and to provide coupling between the elements via evanescent field. Then a narrowing of the emission beam was demonstrated for a two elements array [27]. Later that year, de Naurois et al. developed this buried heterostructure technology, scaled the stripe array comprising up to 16 emitters, while controlling the thermal resistance in a low level [28, 29].

Coupling QCLs by *Y*-junctions is another way to provide strong and robust coupling between sources. Investigating the impacts of coupling length and wavelength variations on the coupling had revealed that the phase-locked emission strongly depends on the length of the

http://dx.doi.org/10.5772/65350

coupling region and also on the ratio between waveguide width and emission wavelength [31, 32]. This understanding of the coupling mechanism is important to exploit *Y*-coupling in mid-infrared QCL arrays with complex structures for high output performance. Then Hoffmann et al. reported a tree-shaped resonator enabling parallel coupling of six laser elements into a single element by means of several *Y*-junctions (see **Figure 7(a)**) [33]. The parallel-coupled branches had three different lengths, which led to complex requirements for the propagating super-modes. When driven far above threshold, all branches could equally emit and the phase was locked throughout the whole device, resulting in a high level of modal control. In-phase emission was observed on both sides of the device (see **Figure 7b** and **c**). However, the measured brightness did not exceed that for single emitters, represented by the low slope efficiency. This was in part attributed to modal competition among branches of different length. To avoid this modal competition, Lyakh et al. introduced a tree array with branches of the same length and the elements merged at the back facet through the *Y*-junctions, as shown in **Figure 8** [34]. This ensured parallel coupling between the elements that entailed in-phase mode dominance and, therefore, led to an on-axis far-field intensity distribution with nearly diffraction-limited divergence. Based on the above observations, one can realize that the *Y*-coupling schemes give high efficient coupling for phase locking, which forms an elementary building block for more complex on-chip interferometric devices.

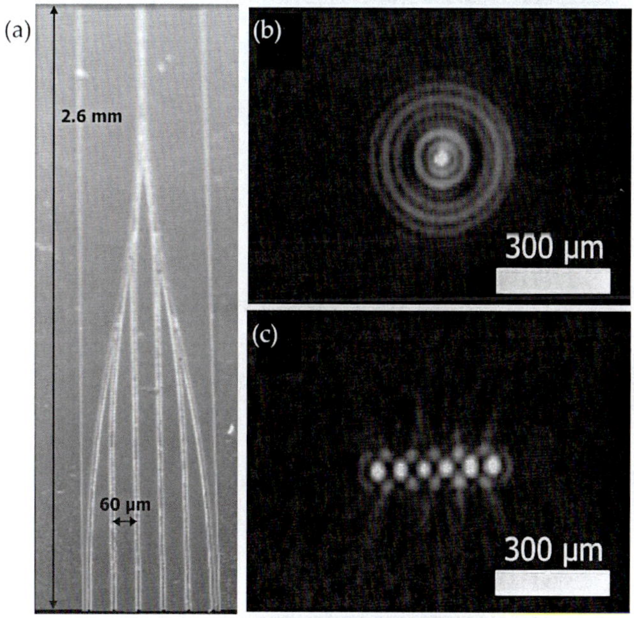

Figure 7. Tilted SEM image of the fabricated tree array QCL (a) with near-field images of the stem facet (b) and the branch facet (c). Citing from Ref. [33].

In integrated diode laser system, leaky-wave coupled devices showed the most robust operation. For mid-infrared QCL systems, by utilizing an anti-guide waveguide structure with high index contrast, Kirch et al. achieved very stable fundamental super-mode QCL array emitting at 8.36 μm for the first time [37]. The leaky-wave coupling schemes showed very high effective coherence between QCL elements, since extremely narrowed centre lobe with most of the total power concentrated have been obtained. Based on this observation, a high-power phase-locked QCL array consisting of 100 elements that were integrated in parallel was achieved at $\lambda \sim 4.6$ μm [38]. With index-guidance and delicate design of gain distribution, an in-phase-like super-mode with a low-divergence beam with an optical power of up to 40 W was obtained. These findings in leaky-wave coupled QCL arrays in the middle-and long-wavelength infrared regions pave the way to further promotion of coupling efficiency and beam quality.

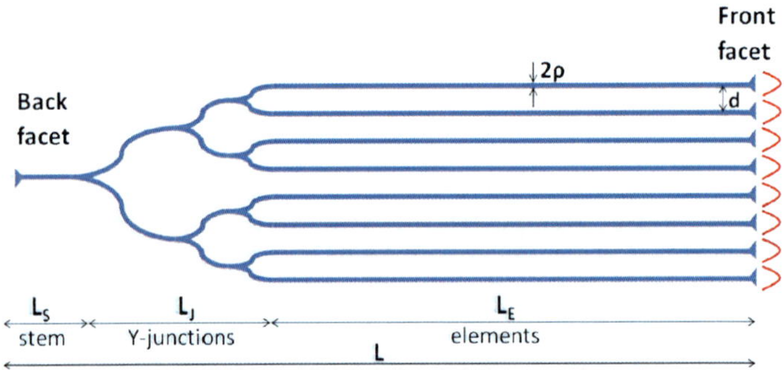

Figure 8. Schematic of the tree-array configuration of QCLs. Citing from Ref. [34].

3.2. Basic consideration for the coupled THz QCLs

Compared with great success of coherent arrays of mid-infrared QCLs, the progress to develop coherent THz QCL arrays is much slower. To effectively and accurately couple THz QCLs is more difficult than mid-infrared ones for several reasons. First, dynamics of the coupled THz QCLs, including optical feedback, mutual injection, modal competition and the effect of gain and loss distributions in the coupled waveguide, have not been clearly understood yet. Second, for THz QCLs with common single-metal waveguide, there are often several coupling mechanisms existing simultaneously. For example, in a typical evanescent-wave coupling structure, the coupling of the optical fields in substrate may also exist, which will disturb the effect of coherent couplings. Third, delicate technique is needed for the fabrication of coupled waveguides to minimizing waveguide loss and optimising the relative phase control, which is very sensitive to the size of sub-wavelength junction.

As a beginning to understand the phase locking in QCL arrays, we investigate the dynamic behaviours of two coupled QCLs to see the locking condition and stability properties of THz QCL arrays. Using the semi-classical laser theory [40], the time evolution of the electric field

$E_j(t)$ of the jth THz QCL cavity when coupled to the field $E_k(t)$ in the kth QCL cavity, assuming no time-delay effects, spatial hole burning, and lateral diffusion of the carriers, is given by

$$\frac{dE_j(t)}{dt} = \{\mathrm{i}[\omega_j + \frac{1}{2}\alpha_j N_{pj} G_{0j} \Delta N_j(t)] + \frac{1}{2}[N_{pj} G_{0j} N_j(t) - \frac{1}{\tau_{pj}}]\} E_j(t) + q_{jk} E_k(t), \quad (1)$$

where $E_j(t)$ is the complex electric field that has been normalized so that $|E_j(t)|^2$ is the photon density; $N_j(t) = N_{3j}(t) - N_{2j}(t)$ is the carrier density difference between the upper and the lower laser levels; $\Delta N_j(t) = N_j(t) - Nth_j$ is the carrier density changes due to light injection from the other QCL; α_j is the line width enhancement factor; G_{0j} is the optical gain coefficient of one single active region period in QCLs; N_{pj} is the number of stages; τ_{pj} is the lifetime of photon in the cavity; $q_{jk} = q_{kj} = q\mathrm{e}^{\mathrm{i}\beta}$ (due to the reversibility of optical path) denotes the coupling efficiency between the two QCLs with modulus $q = c/\tau_c$ and argument β, where c is the coupling coefficient and τ_c is the laser cavity round trip time.

A three-level rate equation model is used to describe the dynamic behaviour of carriers in each individual QCL. The rate equations for population inversion and carriers in the lower laser level are described as

$$\frac{dN_j(t)}{dt} = J_j - \frac{2[N_j(t) + N_{2j}(t)]}{\tau_{3j}} - 2G_{0j} N_j(t) \left|E_j(t)\right|^2 + \frac{N_{2j}(t)}{\tau_{2j}}, \quad (2)$$

$$\frac{dN_{2j}(t)}{dt} = \frac{N_j(t) + N_{2j}(t)}{\tau_{3j}} + G_{0j} N_j(t) \left|E_j(t)\right|^2 - \frac{N_{2j}(t)}{\tau_{2j}}, \quad (3)$$

where J_j is the current injected into the active region divided by electronic charge e and τ_{2j} and τ_{3j} represent the lifetime of lower and upper laser levels, respectively. Eqs. (1)–(3) are coupled nonlinear differential equations, which represent the interaction between the field in the jth QCL and the field injected from the kth QCL.

The complex electric fields in the above equations can be written as

$$E_j(t) = F_j(t) \mathrm{e}^{\mathrm{i}[\omega_j t + \varphi_j(t)]}, \quad (4)$$

where $F_j(t)$ and $\varphi_j(t)$ are the slowly varying envelope and phase of the electric field, ω_j is the angular frequency of the jth QCL without the coupling to other one.

In order to study the possibility and the stability of the phase locking of the two QCLs, we must first determine the stationary solutions of Eqs. (1)–(3). When the QCLs are phase locked, their fields are assumed to have the form as follows:

$$E_{\mathrm{A}}(t) = F_{\mathrm{A}}(t)\mathrm{e}^{\mathrm{i}(\omega t + \varphi_{\mathrm{L}})}, \tag{5}$$

$$E_{\mathrm{B}}(t) = F_{\mathrm{B}}(t)\mathrm{e}^{\mathrm{i}\omega t}, \tag{6}$$

where ω is a lasing frequency of the coupled laser system, and φ_{L} is the locked phase of the field $E_{\mathrm{A}}(t)$ relative to $E_{\mathrm{B}}(t)$. When this two coupled QCL system is phase locked, the amplitudes and phases remain time independent, and φ_{L} is a constant. This means that a necessary condition for phase locking is that the solutions to Eq. (1) should be given by

$$\varphi_{\mathrm{A}}(t) = (\omega - \omega_{\mathrm{A}})t + \varphi_{\mathrm{L}}, \tag{7}$$

$$\varphi_{\mathrm{B}}(t) = (\omega - \omega_{\mathrm{B}})t, \tag{8}$$

Therefore, the rate equations have a steady-state solution:

$$\Delta N_{\mathrm{A}} = -\frac{2F_{\mathrm{B}}q\cos(\beta - \varphi_{\mathrm{L}})}{F_{\mathrm{A}}N_{\mathrm{pA}}G_{0\mathrm{A}}}, \tag{9}$$

$$\Delta N_{\mathrm{B}} = -\frac{2F_{\mathrm{A}}q\cos(\beta + \varphi_{\mathrm{L}})}{F_{\mathrm{B}}N_{\mathrm{pB}}G_{0\mathrm{B}}}, \tag{10}$$

$$\Delta\omega_{\mathrm{A}} = -\frac{F_{\mathrm{B}}}{F_{\mathrm{A}}}q\sqrt{1+\alpha_{\mathrm{A}}^{2}}\sin(\theta_{\mathrm{A}} - \beta + \varphi_{\mathrm{L}}), \tag{11}$$

$$\Delta\omega_{\mathrm{B}} = -\frac{F_{\mathrm{A}}}{F_{\mathrm{B}}}q\sqrt{1+\alpha_{\mathrm{B}}^{2}}\sin(\theta_{\mathrm{B}} - \beta - \varphi_{\mathrm{L}}), \tag{12}$$

$$F_{j}^{2} = \frac{(\tau_{3j} - \tau_{2j})J_{j} - (\Delta N_{j} + N_{\mathrm{th}j})}{\tau_{3j}G_{0j}(\Delta N_{j} + N_{\mathrm{th}j})}, \tag{13}$$

where $\theta_j = \arctan(\alpha_j)$, $\Delta\omega_j = \omega - \omega_j$.

Eqs. (11) and (12) give the allowed frequency detuning between the free running QCLs with which they could be phase locked by the optical coupling. If α_j, τ_{2j}, τ_{3j}, $N_{\mathrm{th}j}$, G_{0j} and $\tau_{\mathrm{p}j}$ are the same for both QCLs, and the coupling between the lasers are weak, one can assume that $F_{\mathrm{A}} = F_{\mathrm{B}}$. Then, the allowed frequency detuning between the unperturbed QCLs can be obtained as:

$$\Delta = \omega_B - \omega_A \approx -2(c/\tau_c)\sqrt{1+\alpha^2}\cos(\theta-\beta)\sin(\varphi_L), \quad (14)$$

which is determined by the coupling strength and phase, the line width enhancement factor and the locked phase. One can see that when $\varphi_L = 0, \pm\pi$, it only allows lasers with $\omega_A = \omega_B$ to be locked. For QCLs with $\omega_A \neq \omega_B$, i.e. $\Delta \neq 0$, it requires both $\sin\varphi_L \neq 0$ and $\cos(\theta-\beta) \neq 0$. This gives the condition for steady-state solution as $\beta \neq \theta \pm \pi/2$.

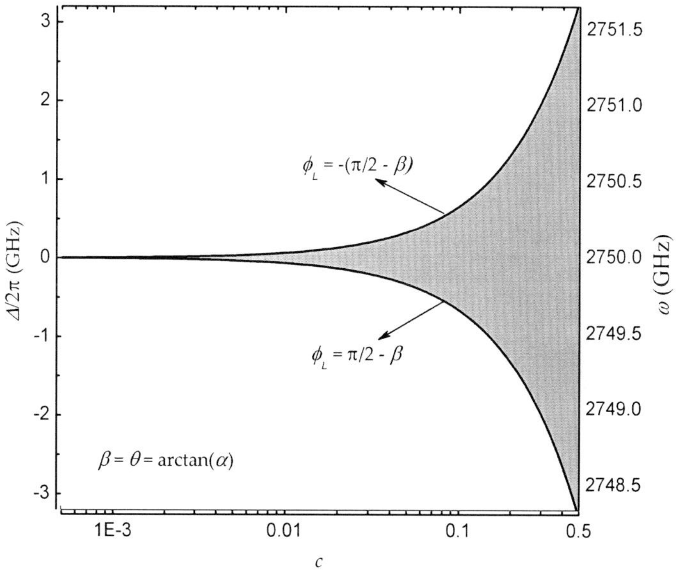

Figure 9. The phase-locking bandwidth of two coupled THz QCLs as a function of the coupling strength.

To find the dynamically stable regime of allowed frequency detuning, one should also perform a small signal analysis on the dynamic equations of the fields and rate equations of carriers. Investigating dynamic behaviours of small perturbations around the stationary solutions for the amplitudes of the fields, the phase difference and the carrier densities, the stability properties of these solutions can be ascertained.

If one of the QCLs' free-running frequency is known, one can calculate the stable frequency for the coupled system, as labelled in the right scale of **Figure 9** for $f_A = \omega_A/2\pi = 2.75$ THz. The stable locking bandwidth varying with the coupling strength is shown in a grey area in **Figure 9**. Here, $\beta = \theta$ and $\alpha = 0.5$ are assumed. The dynamic stability analysis shows that the phase boundaries of the negative and positive detuning edges are $\pi/2-\beta$ and $(\pi/2-\beta)$, which ensure both ΔN_A and ΔN_B are negative in the whole stable regime. Although the calculated stable locking map of QCLs is similar to those of diode lasers [41], the locking range is usually one order of magnitude smaller than that of diode lasers at the same coupling strength. This may be caused by the much smaller line width enhancement factor and longer round trip time for QCLs.

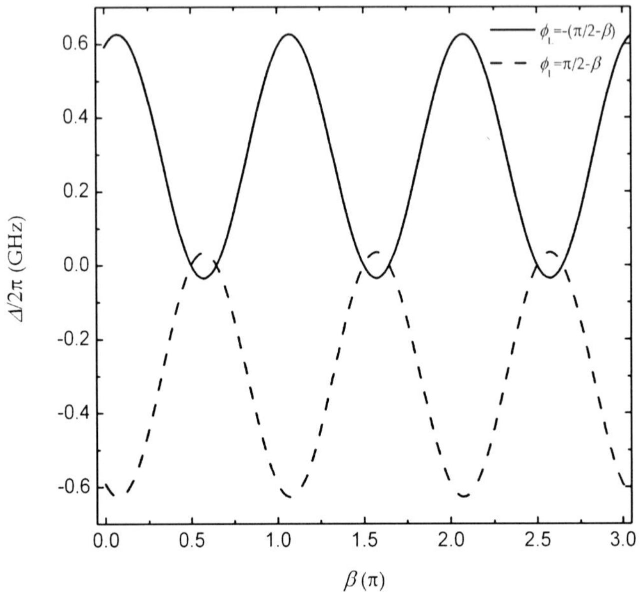

Figure 10. The phase-locking bandwidth of two coupled THz QCLs as functions of the phase of the coupling efficiency β.

Figure 10 gives the dependences of the phase-locked bandwidth on the phase of the coupling efficiency β for $c = 0.1$. It can be found that the stable locking range is modulated by β periodically.

From the dynamic stability analysis of two coupled THz QCLs, one can find a stable phase-locked mode for the coupled system. Operating on this super-mode, the two elements lase with the same frequency and a constant phase difference. Although the analysis shows the possibility of phase-locked lasing of coupled THz QCLs, the phase-locking range is much smaller than that of diode lasers assuming the same coupling strength. This result gives great challenge to design and fabricate the coupling structure for THz QCLs. In practice, to achieve high-quality beam profile, structures of arrays are more complex than that of the two elements array discussed here.

3.3. Developments of phase-locked arrays of THz QCLs

Although phase-locked arrays of THz QCLs experienced relatively slow development, concrete progress has been made during recent few years. The first demonstration of phase-locked arrays of surface-emitting DFB THz QCLs had been reported by Hu's group in 2010

[36]. The surface-emitting DFB array was proposed by connecting element lasers through phase sectors in series, which is schematically shown in **Figure 11(a)**. By choosing a proper phase sector length, the desired in-phase mode would have the lowest surface loss and therefore would be the lasing mode. Up to six laser ridges were locked in-phase with a low-divergent single-lobe far-field beam-pattern (FWHM ≈ 10°) along the array direction (see **Figure 11(b)**).

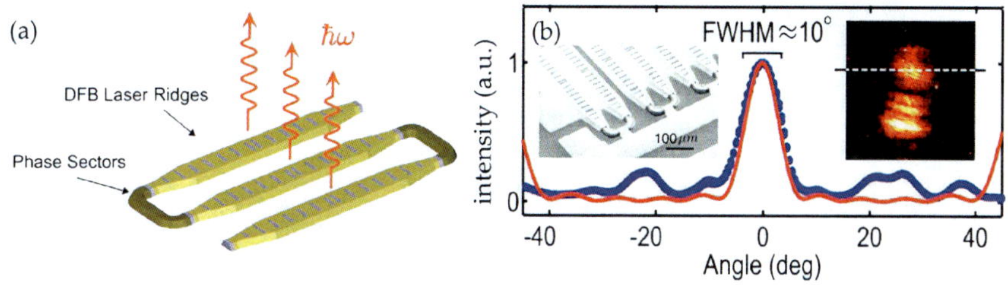

Figure 11. (a) Diagram of a three-ridge surface-emitting DFB array. (b) SEM image of main laser ridges and far-field (20 cm) beam-pattern of the six-ridge array along array direction (x). Citing from Ref. [36].

Very recently, another type of surface-emitting arrays based on second-order DFB THz QCLs has been demonstrated by Halioua et al. [39]. They explored a different coupling scheme combining graded photonic heterostructure THz lasers with a ring resonator, to fix the relative phase (either symmetric or anti-symmetric) between the lasers. Although these surface-emitting arrays achieved dramatic narrowing of the output beam profile, there exists an optimal trade-off between fabrication complexity and output power/beam profile. Further decreasing the beam divergence along both axes requires increasing the radiating aperture. However, if this is done by increasing the width of the waveguide, thermal performance suffers and multiple transverse modes can appear. Therefore, on-chip phase locking is still difficult for large numbers of array elements, and grating side-lobes appear if the array spacing is larger than the wavelength.

A THz vertical-external-cavity surface-emitting-laser (VECSEL) formed by an active meta-surface reflector and a flat output coupler reflector was proposed to address this challenge [30]. Lasing was possible when the meta-surface reflector was placed into a low-loss external cavity such that the external cavity determined the beam properties. A near-Gaussian beam of 4.3° × 5.1° divergence was observed.

Besides the surface-emitting devices, grating-selected single mode can also be achieved in facet-emitting arrays, as demonstrated by Marshall et al. in 2013 [35]. They coupled two THz QCLs in a Y-junction configuration on the same substrate (see **Figure 12**). Frequency control and mode switching were accomplished in one QCL by introducing a holographically designed aperiodic grating. The demonstration of coupling and switchable control in Y-junction coupled THz QCLs illustrated their potential as the basis for future systems, in which on-chip manipulation of spectral information is possible.

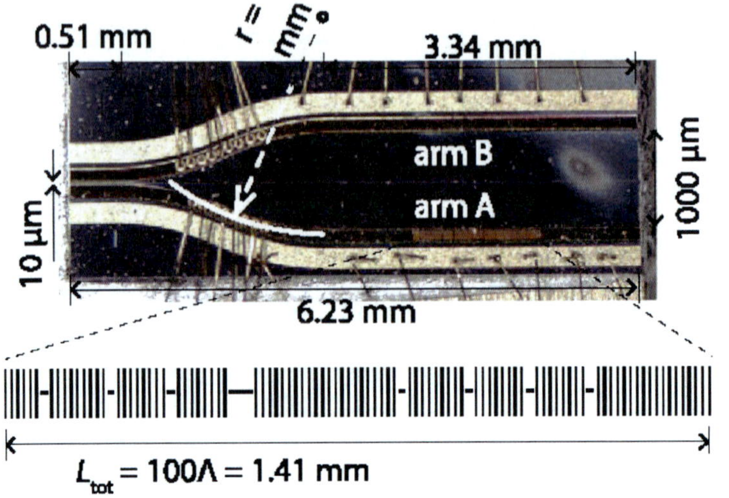

Figure 12. Photograph of Y-coupled THz QCLs, showing the electrical contacts, wire bonding, grating position and critical device dimensions. Citing from Ref. [35].

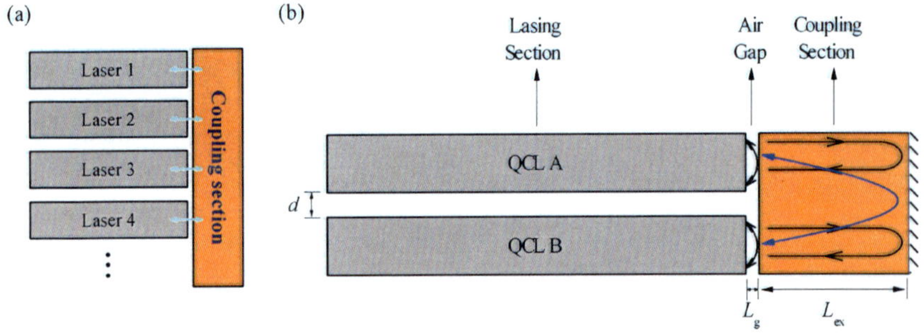

Figure 13. (a) Schematic of the THz QCL array of coupled cavity configuration. (b) Illustration of the mutual injection of the optical fields in two-QCL coupled system.

Compared with surface emission, although the output beam of the facet-emitting devices suffers higher divergence originated from the diffraction of the sub-wavelength size of the aperture, facet-emitting devices are expected to give higher output power due to the longer length of the waveguide along the propagating direction. In 2014, Kundu et al. demonstrated frequency-tunable THz QCL using coupled cavities [42]. In their device, one section of the device (the lasing section) was electrically biased above threshold using a short current pulse, while the other section (the tuning section) was biased below threshold with a wider current pulse to achieve controlled localized electrical heating. This frequency-tunable QCL demon-

http://dx.doi.org/10.5772/65350

strated stability of the device with optical feedback in the coupled cavities. Inspired by this work, we proposed a novel scheme of monolithic THz QCL array based on the similar two-section coupled-cavity. The schematic is shown in **Figure 13**. The array consists of several facet-emitting QCLs (the lasing section) which are electrically biased above threshold. The QCLs are optically coupled via their rear facets by a coupling section, which is biased below threshold. This enables mutual injection of optical fields and also allows the injection strength and phase to be tuned by changing its refraction index. The two sections are optically coupled but electrically isolated through a narrow gap. This compact structure eliminates beam-steering error compensation usually required in non-monolithic external cavity device such as the arrangements in VECSELs, which would be more robust in controlling the beam quality. The theoretical simulation has shown that the array can work in the phase-locked mode and the divergence of the far-field beam will be greatly reduced. Such a monolithic QCL array with mutual injection of optical fields may provide not only a new method of achieving phase-locked arrays, but also a platform for studying complex dynamical behaviours in THz QCLs.

4. Summary

The spectral purity and the output power of a THz QCL, which are key factors for varieties of applications, are contradictory to a certain extent. The optical power of THz QCLs can reach the level of Watt, but they are usually multi-mode operated. The power amplification and coherent combination techniques are important ways to improve the output power maintaining the single-mode operation. Meanwhile, they are also helpful to improve the beam quality of THz QCLs. For power amplification, the tapered active region has been demonstrated as a convenient way to improve the output power of THz QCLs and both surface and lateral grating techniques have been employed to ensure the single-mode operation. For coherent power combination, the great success has been made in diode lasers and mid-infrared QCLs. The simulation of the dynamics of coupled THz QCLs has revealed the basic difficulty of phase-locking operation. Nevertheless, with elaborate designs of coupling structure, the coherent arrays of surface-emitting and facet-emitting THz QCLs have been demonstrated. We also present a new design of monolithic THz QCLs array with optical mutual injection and show the possibility of coherent lasing and the reduction for the far-field beam. The developments of power amplification and coherent combination techniques will greatly promote the practicability of THz QCLs especially in the fields where the spectral purity is desired.

Acknowledgements

Financial supports from the National Basic Research Program of China (973 Program, Grant nos. 2013CB632805) and the Foundation of Director of IAPCM (ZYSZ1518-16) are gratefully acknowledged.

Author details

Yan Xie[1], Yanfang Li[1], Jian Wang[2], Ning Yang[1*], Weidong Chu[1] and Suqing Duan[1]

*Address all correspondence to: yang_ning@iapcm.ac.cn

1 Institute of Applied Physics and Computational Mathematics, Beijing, China

2 Department of Physics, Beijing Jiaotong University, Beijing, China

References

[1] Köhler R., Tredicucci A., Beltram F., Beere H. E., Linfield E. H., Davies A. G., et al. Terahertz semiconductor-heterostructure laser. Nature. 2002;417:156. DOI:10.1038/417156a

[2] Faist J., Capasso F., Sivco D. L., Sirtori C., Hutchinson A. L., and Cho A. Y. Quantum cascade laser. Science. 1994;264:553–556. DOI: 10.1126/science.264.5158.553

[3] Williams B. S. Terahertz quantum-cascade lasers. Nat. Photonics. 2007;1:517–525. DOI: 10.1038/nphoton.2007.166

[4] Rauter P. and Capasso F. Multi-wavelength quantum cascade laser arrays. Laser Photon. Rev. 2015;9(5):452–477. DOI: 10.1002/lpor.201500095

[5] Walpole J. N. Semiconductor amplifiers and lasers with tapered gain regions. Opt. Quant. Electron. 1996;28(6):623–645. DOI 10.1007/BF00411298

[6] Pfahler C., Eichhorn M., Keleman M. T., Mikulla M., Schmitz J., and Wagner J. Gain saturation and high-power pulsed operation of GaSb-based tapered diode lasers with separately contacted ridge and tapered section. Appl. Phys. Lett. 2006;89(2):021107. DOI: 10.1063/1.2218823

[7] Wenzel H., Paschke K., Brox O., Bugge F., Fricke J., Ginolas A., et al. 10 W continuous-wave monolithically integrated master-oscillator power-amplifier. Electron. Lett. 2007;43(3):160–162. DOI: 10.1049/EL:20073297

[8] Vijayakumar D., Jensen O. B., Ostendorf R., Westphalen T., and Thestrup B. Spectral beam combining of a 980 nm tapered laser bar. Opt. Express. 2010;18(2):893–898. DOI: 10.1364/OE.18.000893

[9] Nähle L., Semmel J., Kaiser W., Höfling S., and Forchel A. Tapered quantum cascade lasers. Appl. Phys. Lett. 2007;91(18):181122. DOI: 10.1063/1.2805628

[10] Menzel S., Diehl L., Pflügl C., Goyal A., Wang C., Sanchez A., et. al. Quantum cascade laser master-oscillator power-amplifier with 1.5 W output power at 300 K. Opt. Express. 2011;19(17):16229–16235. DOI: 10.1364/OE.19.016229

[11] Lyakh A., Maulini R., Tsekoun A., Go R., and Patel C. K. N. Tapered 4.7 μm quantum cascade lasers with highly strained active region composition delivering over 4.5 watts of continuous wave optical power. Opt. Express. 2012;20(4):4382–4388. DOI: 10.1364/OE.20.004382

[12] Rauter P., Menzel S., Goyal A. K., Gökden B., Wang C. A., Sanchez A., et. al. Master-oscillator power-amplifier quantum cascade laser array. Appl. Phys. Lett. 2012;101(26): 161117. DOI: 10.1063/1.4773377

[13] Kirch J. D., Shin J. C., Chang C.-C., Mawst L. J., Botez D., and Earles T. Tapered active-region quantum cascade lasers (λ = 4.8 μm) for virtual suppression of carrier-leakage currents. Electron. Lett. 2012;48(4):234–235. DOI: 10.1049/el. 2012.0017

[14] Li Y., Wang J., Yang N., Liu J., Wang T., Liu F., et al. The output power and beam divergence behaviors of tapered terahertz quantum cascade lasers. Opt. Express. 2013;21(13):15998–16006. DOI: 10.1364/OE.21.015998

[15] Maker P. D. and Terhune R.W. Study of optical effects due to an induced polarization third order in the electric field strength. Phys. Rev. 1965;137(3A): 801–818. DOI:10.1103/PhysRev.137.A801

[16] Wang C. C. Length-dependent threshold for stimulated Raman effect and self-focusing of laser beams in liquids. Phys. Rev. Lett. 1966;16(9):344. DOI:10.1103/ PhysRevLett.16.344

[17] Wang J., Li Y., and Yang N. The origin of self focusing effect in terahertz quantum cascade lasers. J. Phys. Conf. Ser. 2015;574:012053. DOI: 10.1088/1742-6596/574/1/012053

[18] Wang T., Liu J., Chen J., Liu F., Wang L., et al. High-power distributed feedback terahertz quantum cascade lasers. IEEE Electron Device Lett. 2013;34(11):1412–1414. DOI: 10.1109/LED.2013.2280713

[19] Wang T., Liu J., Liu F., Wang L., Zhang J., and Wang Z. Tri-channel single-mode terahertz quantum cascade laser. Opt. Lett. 2014;39(23):6612–6615. DOI: 10.1364/OL.39.006612

[20] Yao C., Xu T.H., Wan W.J., Li H., and Cao J. C. Single-mode tapered terahertz quantum cascade lasers with lateral gratings. Solid-State electron. 2016;122:52–55. DOI: 10.1016/j.sse.2016.04.008

[21] Lee B. G, Belkin M. A., Audet R., MacArthur J., Diehl L., Pflügl C., et al. Widely tunable single-mode quantum cascade laser source for mid-infrared spectroscopy. Appl. Phys. Lett. 2007;91(23):231101. DOI: 10.1063/1.2816909

[22] Mujagić E., Hoffmann L. K., Schartner S., Nobile M., Schrenk W., Semtsiv M. P., et al. Low-divergence single-mode surface emitting quantum cascade ring lasers. Appl. Phys. Lett. 2008;93(16):161101. DOI: 10.1063/1.3000630

[23] Mujagić E., Schwarzer C., Yao Y., Chen J., Gmachl C., and Strasser G. Two-dimensional broadband distributed-feedback quantum cascade laser arrays. Appl. Phys. Lett.. 2011;98(14):141101. DOI: 10.1063/1.3574555

[24] Rauter P., Menzel S., Goyal A. K., Wang C. A., Sanchez A., Turner G., et al. High-power arrays of quantum cascade laser master-oscillator power-amplifiers. Opt. Expr. 2013;21(4):4518. DOI: 10.1364/OE.21.004518

[25] Lee B. G., Kansky J., Goyal A. K., Pflügl C., Diehl L., Belkin M. A., et al. Beam combining of quantum cascade laser arrays. Opt. Express. 2009;17(18):16216. DOI: 10.1364/OE. 17.016216

[26] Jiang A., Jung S., Jiang Y., Vijayraghavan K., Kim J. H., and Belkin M. A. Widely tunable terahertz source based on intra-cavity frequency mixing in quantum cascade laser arrays. Appl. Phys. Lett.. 2015;106(26):261107. DOI: 10.1063/1.4923374

[27] Bismuto A., Amanti M., Beck M., and Faist J. Buried-heterostructure phase-locked arrays of mid-infrared quantum cascade lasers. In: Science and Innovations 2011; 1–6 May 2011; Baltimore, Maryland, United States. Optical Society of America; 2011. DOI: 10.1364/CLEO_SI.2011.CTuV2

[28] de Naurois G. M., Carras M.,Simozrag B., Patard O., Alexandre F., and Marcadet X. Coherent quantum cascade laser micro-stripe arrays. AIP Adv. 2011;1(3):032165. DOI: 10.1063/1.3643690

[29] de Naurois G. M., Simozrag B., Maisons G., Trinité V., Alexandre F., and Carras M. High thermal performance of μ-stripes quantum cascade laser. Appl. Phys. Lett.. 2012;101(4): 041113. DOI: 10.1063/1.4739004

[30] Xu L., Curwen C. A., Hon P. W. C., Chen Q.-S., Itoh T., and Williams B. S. Metasurface external cavity laser. Appl. Phys. Lett. 2015;107(22):221105. DOI: 10.1063/1.4936887

[31] Hoffmann L. K., Hurni C. A., Schartner S., Austerer M., Mujagić E., Nobile M., et al. Coherence in Y-coupled quantum cascade lasers. Appl. Phys. Lett.. 2007;91(16):161106. DOI: 10.1063/1.2800293

[32] Hoffmann L. K., Hurni C. A., Schartner S., Mujagić E., Andrews A. M., Klang P., et al. Wavelength dependent phase locking in quantum cascade laser Y-junctions. Appl. Phys. Lett.. 2008;92(6):061110. DOI: 10.1063/1.2841634

[33] Hoffmann L. K., Klinkmüller M., Mujagić E., Semtsiv M. P., Schrenk W., Masselink W. T., et al. Tree array quantum cascade laser. Opt. Express. 2009;17(2):649–657. DOI: 10.1364/OE.17.000649

[34] Lyakh A., Maulini R., Tsekoun A., Go R., and Patel C. K. N. Continuous wave operation of buried heterostructure 4.6μm quantum cascade laser Y-junctions and tree arrays. Opt. Express. 2014;22(1):1203–1208. DOI: 10.1364/OE.22.001203

[35] Marshall O.P., Chakraborty S., Khairuzzaman M., Beere H. E., and Ritchie D. A. Reversible mode switching in Y-coupled terahertz lasers. Appl. Phys. Lett. 2013;102(11): 111105. DOI: 10.1063/1.4796039

[36] Kao T.-Y., Hu Q., and Reno J. L. Phase-locked arrays of surface-emitting terahertz quantum-cascade lasers. Appl. Phys. Lett. 2010;96(10):101106. DOI: 10.1063/1.3358134

[37] Kirch J. D., Chang C.-C., Boyle C., Mawst L. J., Lindberg III D., Earles T., et al. 5.5 W near-diffraction-limited power from resonant leaky-wave coupled phase-locked arrays of quantum cascade lasers. Appl. Phys. Lett. 2015;106(6):061113. DOI: 10.1063/1.4908178

[38] Yan F.-L., Zhang J.-C., Jia Z.-W., Zhuo N., Zhai S.-Q., Liu S.-M., et al. High-power phase-locked quantum cascade laser array emitting at λ ~ 4.6 μm. AIP Adv. 2016;6(3):035022. DOI: 10.1063/1.4945383

[39] Halioua Y., Xu G., Moumdji S., Li L., Zhu J., Linfield E. H., et al. Phase-locked arrays of surface-emitting graded-photonic-heterostructure terahertz semiconductor lasers. Opt. Express. 2015;23(5):6915. DOI: 10.1364/OE.23.006915

[40] Spencer M. and Lamb W. Theory of Two Coupled Lasers. Phys. Rev. A. 1972;5(2):893. DOI: 10.1103/PhysRevA.5.893

[41] Tsacoyeanes J. G. Phase locking and stability properties for two coupled semiconductor lasers. J. Appl. Phys. 1988;64(1):32. DOI: 10.1063/1.341431

[42] Kundu J. Dean P., Valavanis A., Chen L., Li L., Cunningham J. E., et al. Discrete Vernier tuning in terahertz quantum cascade lasers using coupled cavities. Opt. Express. 2014;22(13):16595. DOI: 10.1364/OE.22.016595